Mechatronics Sourcebook

Newton C. Braga

THOMSON

DELMAR LEARNING   Australia   Canada   Mexico   Singapore   Spain   United Kingdom   United States

THOMSON
★
DELMAR LEARNING

Mechatronics Sourcebook

Newton C. Braga

Executive Director:
Alar Elken

Editorial Assistant:
Jennifer Luck

Executive Production Manager:
Mary Ellen Black

Executive Editor:
Sandy Clark

Executive Marketing Manager:
Maura Theriault

Production Manager:
Andrew Crouth

Acquisitions Editor:
Gregory L. Clayton

Marketing Coordinator:
Sarena Douglass

Production Coordinator:
Dawn Jacobson

Library of Congress Cataloging-in-Publication Data:
Daugherty, Dawn

Mechatronics sourcebook /
Newton C. Braga

p.   cm.

ISBN 1-40181-432-8
1. Mechatronics   I. Title
   TJ163.12 .B73 2003
   621—dc21
                                    2002014298

This book is printed on acid-free paper.

NOTICE TO THE READER

Preface

Some years ago, when you intended to install a new machine in your plant, you had to call in professionals from three different disciplines: a mechanical engineer who installed the machine in the correct location, adjusting and choosing the correct position of equipment and mechanical components; the electronic or electrical engineer who installed the electronic controls and power supplies; and the software specialist who programmed the microcontrollers or microcomputer used to manage machine operation. Today, things have changed, and you have to call only one professional—a mechatronics specialist who knows how to install the machine, assemble the electronic circuits, and create and install the control software.

Mechatronics is the result of a union of several sciences, including electronics, mechanics, pneumatics, hydraulics, and others, and within the mechatronics category we have such divisions as robotics, pneutronics, hydrotronics, artificial intelligence, and so on. The subject is in the spotlight today, and there is a need for information not only for the professional but also for students, in the form of basic textbooks, courses, and general references.

In my home country of Brazil, I am presently employed in the production of two mechatronics magazines and have created an experimental course on the subject that will be taught in secondary schools. In my work with the magazines, I am in contact with many local universities and industries and have access to material generated by many teachers and researchers.

The general idea is that the field of mechatronics addresses one of the primary purposes of secondary schools: to teach technology, using a practical approach, that (a) induces students to think about their future and (b) teaches skills and vocations. But, to teach practical mechatronics or its basic principles, a great deal of general information, derived from many fields, is necessary. We need to include information from physics, mathematics, electronics, mechanics, computer science, chemistry, and many other subjects.

So, my idea when starting this book was just to assemble and organize the store of information that I have accumulated about all these subjects over the past two years. Much of the information is derived from book I published last year, which includes more that 400 practical circuits for mechatronic and robotic applications.

The present volume is intended as a reference book or sourcebook for mechatronics designers and professionals, students, and hobbyists. It includes useful formulas, tables, practical ideas and information, circuits, application examples, tables of component characteristics, and basically everything professionals, students, and teachers will need in their day-to-day activities related to the subject.

This is not an exhaustive book (I have no such pretension), but it contains what I believe to be essential and often hard-to-find information that can be used on a daily basis. Given unlimited time, I could have included much more, but this could easily become an endless task. Each additional piece of information logically leads to another and then another.

Ultimately, if I can offer the reader a set of fundamentals to simplify his work in the field of mechatronics, perhaps I will be considered to be a fully competent author.

Newton C. Braga

Acknowledgments

I would like to thank all of the people who contributed to making this book possible. In particular, I would like to thank my agent, Jeff Eckert, who has helped not only in that capacity but as a co-producer, offering ideas for changes in the text, adding some footnotes, and having the patience to handle all my confused originals.

Contents

Chapter 17

Appendix A

CHAPTER 1

Mathematics, Physics, and Chemistry: Formulas and Tables

MATHEMATICS

When working on a mechatronics project, mathematical calculation is always necessary. In this chapter, the reader will find useful information about mathematics as applied to mechatronics, including formulas, tables, common values of constants, coordinates changes, binary numbers, and so on.

Table 1.1   Important Logarithms

Log	Value	Log	Value
log 1	0	log 0.001	−3
log 10	1	ln a	$2.3026 \times \log a$
log 100	2	ln 10	2.3026
log 1000	3	log a	$0.4343 \times \ln a$
log 0.1	−1	log e	0.03443
log 0.01	−2		

Table 1.2   Values of π

π	3.141592	$\pi/2$	1.570796
2π	6.283185	$\pi/3$	1.047197
3π	9.424779	$\pi/4$	0.785398
4π	12.566379	$\pi/5$	0.628318
5π	15.707963	$4\pi/3$	4.188790

Table 1.2   Values of π (continued)

6π	18.849556	$4/\pi$	1.273239
7π	21.991486	$3/\pi$	0.954929
8π	25.132741	π^2	9.869604
9π	28.274334	$3\sqrt{\pi}$	1.464592
10π	31.415926	$\log \pi$	0.497715
$1/\pi$	0.318310	$\log \pi$	0.798180
$1/2\pi$	0.159155		

Table 1.3   Relation between Circular and Angular Measures[*]

Degrees	Radians	Radians	Degrees	Radians	Radians
0	0	0	90	$\pi/2$	1.57
15	$\pi/12$	0.26	120	$2\pi/3$	2.10
30	$\pi/6$	0.52	150	$5\pi/6$	2.62
45	$\pi/4$	0.79	180	π	3.14
60	$\pi/3$	1.05	270	$3\pi/2$	4.71
75	$5\pi/12$	1.31	360	2π	6.28

[*] 360 degrees = 2 π radians, 1 radian = 57.2958 degrees.

BINARY TO DECIMAL CONVERSION

Digital circuits use a base 2 system. This means that only two digits are used to represent any quantity. Converting a pure digital number to an equivalent decimal (base 10) can be accomplished using Eq. (1.1) (see also Fig. 1.1).

Formula: Pure binary to BCD conversion

$$Dn = b_1 x_1^0 + b_2 x_2^1 + b_3 x_2^2 + \ldots + b_{n-1} x_2^n \tag{1.1}$$

where Dn = the decimal number

b_1 = the least significant bit (LSB) of the binary number

b_2 to b_{n-1} = the intermediate bits of the binary number

b_n = the most significant digit (MSB) of the binary number

2^0 to 2^n = power of two (see Table 1.4)

BYTE TO DECIMAL CONVERSION

The byte is an eight-bit binary number. Equation (1.2) can be used to convert a byte into decimal form (see Table 1.5).

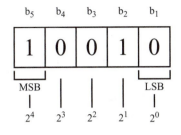

Figure 1.1

Table 1.4   Powers of 2

Power of 2	Decimal	Power of 2	Decimal	Power of 2	Decimal
2^0	1	2^{11}	2,048	2^{22}	4,194,304
2^1	2	2^{12}	4,096	2^{23}	8,388,608
2^2	4	2^{13}	8,192	2^{24}	16,777,216
2^3	8	2^{14}	16,384	2^{25}	33,554,432
2^4	16	2^{15}	32,768	2^{26}	67,108,864
2^5	32	2^{16}	65,536	2^{27}	134,217,728
2^6	64	2^{17}	131,072	2^{28}	268,435,456
2^7	128	2^{18}	262,144	2^{29}	536,870,912
2^8	256	2^{19}	524,288	2^{30}	1,073,741,824
2^9	512	2^{20}	1,048,576	2^{31}	2,147,483,648
2^{10}	1,024	2^{21}	2,097,152	2^{32}	4,294,967,296

Formula: *Byte to decimal conversion*

$$Dn = b_1 x_2^0 + b_2 x_2^1 + b_3 x_2^2 + b_4 x_2^3 + b_5 x_2^4 + b_6 x_2^5 + b_7 x_2^6 + b_8 x_2^7 \qquad (1.2)$$

where Dn = the decimal number

b_1 to b_8 = the bits of the byte

b_1 = the MSB (most significant bit)

b_8 = the LSB (least significant bit)

Table 1.5   Decimal Integers to Pure Binaries

Decimal integer	Binary	Decimal integer	Binary
00	00000000	36	00100100
01	00000001	37	00100101
02	00000010	38	00100110
03	00000011	39	00100111
04	00000100	40	00101000
05	00000101	41	00101001
06	00000110	42	00101010
07	00000111	43	00101011
08	00001000	44	00101100
09	00001001	45	00101101
10	00001010	46	00101110
11	00001011	47	00100111
12	00001100	48	00110000
13	00001101	49	00110001
14	00001110	50	00110010
15	00001111	51	00110011
16	00010000	52	00110100
17	00010001	53	00110101
18	00010010	54	00110110
19	00010011	55	00110111
20	00010100	56	00111000
21	00010101	57	00111001
22	00010110	58	00111010
23	00010111	59	00111011
24	00011000	60	00111100
25	00011001	61	00111101
26	00011010	62	00111110
27	00011011	63	00111111
28	00011100	64	01000000
29	00011101	65	01000001
30	00011110	66	01000010
31	00011111	67	01000011
32	00100000	68	01000100
33	00100001	69	01000101
34	00100010	70	01000110
35	00100011		

BCD TO DECIMAL CONVERSION

Binary-coded decimal (BCD) is a form of binary representation used in digital electronics. In this system, groups of four bits represent a decimal digit as shown in Fig. 1.2. In the figure, the left-hand group represents a 6 in the tens column, and the right-hand group represents a 5 in the ones column, thereby giving the decimal number 65. Conversion to decimal is made as follows:

Formula: BCD to decimal conversion

$$D_d = b_1 \times 2^0 + b_2 \times 2^1 + b_3 \times 2^2 + b_3 \times 2^3 \tag{1.3}$$

and

$$D_d < 10$$

where D_d = the decimal digit
b_1 to b_4 = the BCD digits or bits
b_1 = the MSB (most significant bit)
b_4 = the LSB (least significant bit)

NEGATIVE POWERS OF TWO

Negative powers of two are shown in Table 1.6.

HEXADECIMAL TO DECIMAL CONVERSION

In the hexadecimal numbering system, digits from 0 through 9 are used with the addition of letters from A through F. As in the case of binary and decimal numbers, the value of a hexadecimal number depends on its horizontal position. Conversion to decimal is accomplished using Equation 1.4. The Table 1.7 gives the values of each digit in the hexadecimal numbering system, and Table 1.8 shows the exponent values of 16.

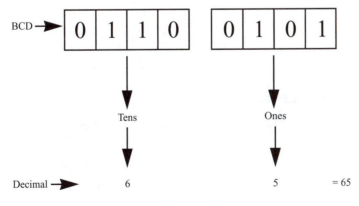

Figure 1.2

Table 1.6   Negative Powers of Two

Negative power of two	Decimal	Negative power of two	Decimal
2^0	1	2^{-9}	0.001 953 125
2^{-1}	0.5	2^{-10}	0.000 976 562 5
2^{-2}	0.25	2^{-11}	0.000 488 281 25
2^{-3}	0.125	2^{-12}	0.000 244 140 625
2^{-4}	0.0625	2^{-13}	0.000 122 070 312 5
2^{-5}	0.031 25	2^{-14}	0.000 061 035 156 25
2^{-6}	0.015 625	2^{-15}	0.000 030 517 578 125
2^{-7}	0.007 812 5	2^{-16}	0.000 015 258 789 062 5
2^{-8}	0.003 906 25		

Formula: *Hexadecimal to decimal conversion*

$$Dn = h_1 \times 16^2 + h_2 \times 16^1 + h_3 \times 16^2 + h_4 \times 16^4 \tag{1.4}$$

where Dn = the decimal number
h_1 to h_4 = the hexadecimal digits
h_1 = the LSB hexadecimal digit
h_4 = the MSB hexadecimal digit

Table 1.7   Hexadecimal Digits and Corresponding Decimals

Hexadecimal	Decimal	Hexadecimal	Decimal
0	0	8	8
1	1	9	9
2	2	A	10
3	3	B	11
4	4	C	12
5	5	D	13
6	6	E	14
7	7	F	15

Table 1.8   Powers of 16

Power of 16	Decimal	Power of 16	Decimal
16^0	1	16^4	65,536
16^1	16	16^5	1,048,576
16^2	256	16^6	16,777,216
16^3	4096		

DECIMAL TO BINARY CONVERSION

There is no formula to convert a decimal number to pure binary. To do so, we have to use an algorithm consisting of successive divisions of the decimal number by the binary base 2. The use of this algorithm is shown in the following text.

Algorithm: *Decimal to binary conversion*

$$b_1 \ldots b_n = \left[\frac{Dn}{2}\right]^{\sigma} R \tag{1.5}$$

The binary number is found by writing, in reverse order, the rest of the successive divisions of the decimal number by two, beginning with the result of the last division.

where b_1 to b_1 = the binary number

Dn = the decimal number

R = the result of the divisions in inverse order

CONVERSION OF THREE-DIMENSIONAL POLAR TO THREE-DIMENSIONAL CARTESIAN COORDINATES

This conversion is illustrated in Fig. 1.3.

Formula:

$$(\phi, \theta, r) \Rightarrow (x, y, z) \tag{1.6a}$$

$$x = r\cos\phi \cdot \cos\theta \tag{1.6b}$$

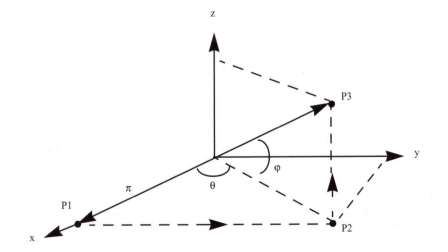

Figure 1.3

$$y = r\cos\phi \cdot \sin\theta \qquad (1.6c)$$

$$z = r\sin\theta \qquad (1.6d)$$

CONVERSION OF THREE-DIMENSIONAL CYLINDRICAL TO THREE-DIMENSIONAL CARTESIAN COORDINATES

This conversion is illustrated in Fig. 1.4.

Formula:

$$(\theta, y, z) \Rightarrow (x, y, z) \qquad (1.7a)$$

$$x = r\cos\theta \qquad (1.7b)$$

$$y = r\sin\theta \qquad (1.7c)$$

$$z = z \qquad (1.7d)$$

CONVERSION OF THREE-DIMENSIONAL SPHERICAL TO THREE-DIMENSIONAL CARTESIAN COORDINATES

This conversion is illustrated in Fig. 1.5.

Formula:

$$(\theta, \varphi, \rho) \Rightarrow (x, y, z) \qquad (1.8a)$$

$$x = \rho\cos\varphi\cos\theta \qquad (1.8b)$$

$$y = \rho\cos\varphi\sin\theta \qquad (1.8c)$$

$$z = \rho\sin\varphi \qquad (1.8d)$$

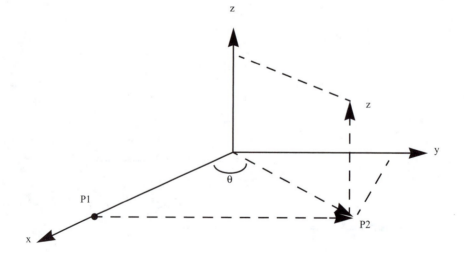

Figure 1.4

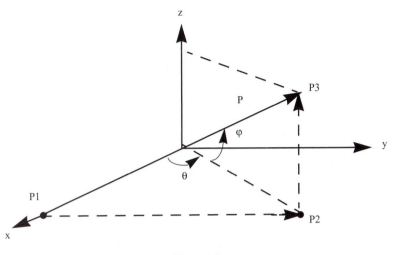

Figure 1.5

GENERAL PHYSICS

UNITS

The following units are based in the International System of Units (*Système International d'Unités),* abbreviated SI, adopted at the *Conférence Générale des Poids et Mesures* in 1960. The SI uses the basic units shown in Table 1.9.

Table 1.9   SI Units

Base quantity	Name	Symbol
Amount of substance	mole	mol
Electric current	ampere	A
Length	meter	m
Luminous intensity	candela	cd
Mass	kilogram	kg
Thermodynamic temperature	Kelvin	K
Time	second	s

Derived quantity	Name	Symbol
Acceleration	meters per second per second	m/s^2
Area	square meter	m^2
Current density	amperes per square meter	A/m^2
Luminance	candelas per square meter	cd/m^2
Magnetic field strength	amperes per meter	A/m

Table 1.9   SI Units *(continued)*

Base quantity	Name	Symbol
Mass fraction	kilograms per kilogram*	1
Specific volume	cubic meters per kilogram	m^3/kg
Speed or velocity	meters per second	m/s
Substance concentration	moles per cubic meter	mol/m^3
Volume	cubic meter	m^3
Wave number	reciprocal meter	m^{-1}

* Represented by the numeral 1.

OTHER UNITS NOT INCLUDED IN THE SI

The units shown in Table 1.10 are acceptable for use with SI units.

Table 1.10   Other Units

Name	Value in SI units	Symbol
Angstrom	$1 \text{ Å} = 0.1 \text{ nm} = 10^{-10} \text{ m}$	Å
Are	$1 \text{ a} = 100 \text{ m}^2$	a
Astronomical unit	$1 \text{ ua} = 1.495\ 98 \times 10^{11}$	ua
Bar	$1 \text{ bar} = 1 \text{ Mpa} = 100 \text{ kPa} = 1000 \text{ Pa}$	bar
Barn	$1 \text{ b} = 100 \text{ fm}^2 = 10^{-28} \text{ m}^2$	b
Bel	$1 \text{ B} = 1/2 \ln 10 \text{ Np}$	B
Curie	$1 \text{ ci} = 3.7 \times 10^{10} \text{ Bq}$	Ci
Degree (angle)	$1° = \pi/180 \text{ rad}$	°
Hectare	$1 \text{ ha} = 10{,}000 \text{ m}^2$	ha
Hour	$1 \text{ hour} = 60 \text{ min} = 3600 \text{ s}$	h
Knot	1 nautical mile per hour = 1852/3600 m/s	–
Liter	$1 \text{ l} = \text{dm}^3$	l
Minute	$1 \text{ min} = 60 \text{ s}$	min
Minute (angle)	$1' = 1/60°$	′
Nautical mile	1852 m	–
Neper	$1 \text{ Np} = 1$	Np
Rad	$1 \text{ rad} = 1 \text{ cGy} = 10^{-2} \text{ Gy}$	rad
Rem	$1 \text{ rem} = 1 \text{ cSv} = 10^{-2} \text{ Sv}$	rem
Roentgen	$1 \text{ R} = 2.58 \times 10^{-4} \text{ C/kg}$	R
Second (angle)	$1'' = 1/60'$	″
Square mil	$1 \text{ sq mil} = 645.2 \text{ mm}^2$	sq mil

MISCELLANEOUS UNIT CONVERSIONS[*]

1 atmosphere	=	101,325 pascals
1 bar	=	100,000 pascals
1 BTU/cubic foot	=	9547 kcal/cubic meter = 39 964 kilojoule/cubic meter
1 centpoise	=	0.001 newton-second/square meter
1 chain	=	22 yards = 20.11 meters
1 English mile	=	1760 yards = 1609 meters
1 fluid ounce (U.S.)	=	$2.95735295625 \times 10^3$ cubic meter
1 foot	=	0.3048 meter
1 gallon (U.S.)	=	3.785411784 liters
1 geographical mile	=	7420 meters
1 gram	=	0.001 kilogram
1 inch	=	0.0254 meter
1 liter	=	0.001 cubic meter
1 micron	=	0.000 001 meter
1 mil	=	0.001 in = 0.0254 mm
1 nautical mile	=	1852 meters
1 ounce mass (avdp)	=	0.028349523125 kilogram
1 pascal	=	1 newton/square meter
1 pound force (avdp)	=	4.4482216152605 newtons
1 pound mass (avdp)	=	0.45359237 kilogram
1 quarter (GB)	=	28 lbf = 124.5 kN
1 stone (GB)	=	14 lbf = 62.3 kN
1 yard	=	3 feet = 0.914 meters

SI PREFIXES

Table 1.11 SI Prefixes

Factor	Prefix	Symbol	Factor	Prefix	Symbol
10^{18}	exa	E	10^{-1}	deci	d
10^{15}	peta	P	10^{-2}	centi	c
10^{12}	tera	T	10^{-3}	milli	m
10^{9}	giga	G	10^{-6}	micro	μ
10^{6}	mega	M	10^{-9}	nano	n
10^{3}	kilo	k	10^{-12}	pico	p
10^{2}	hecto	h	10^{-15}	femto	f
10	deca	da	10^{-18}	atto	a

[*]According to the U.S. National Institute of Standards and Technology, www.nist.gov.

CGS UNITS

For mechanical units, the CGSE and CGSM systems coincide fully. The fundamental units of these systems are the centimeter (c), gram (g), and second (s).

Table 1.12　CGS Units

Quantity	Name	Symbol
Acceleration	centimeters per second squared	cm/s^2
Energy and work	erg	erg
Force	dyne	dyn
Length	centimeter	cm
Mass	gram	g
Time	second	s (sec)
Velocity	centimeters per second	cm/s

GREEK ALPHABET

Table 1.13　Greek Alphabet

Upper case	Lower case	Name	Common uses
A	α	alpha	*Lower case:* Transistor current gain (common base); angular acceleration; linear coefficient of expansion
B	β	beta	Transistor current gain (common emitter)
Γ	γ	gamma	*Lower case:* angle; electrical conductivity; Euler constant; gamma ray; pressure coefficient; propagation constant; surface tension; proton gyromagnetic ratio; cubic coeeficient of expansion. *Upper case:* complex Hertzian vector; gamma function; reciprocal inductance
Δ	δ	delta	Increment or decrement of a quantity
E	ε	epsilon	*Lower case:* eltric intensity; permitivity; natural log base
Z	ζ	zeta	displacement component of a sound-bearing particle
H	η	eta	*Lower case:* efficiency; electric suceptibility; dynamic viscosity
Θ	θ	theta	*Lower case:* angle; phase displacement; thermal resistance *Upper case:* angle; image transfer constant; phase angle
I	ι	iota	–
K	κ	kappa	–
Λ	λ	lambda	*Lower case:* wavelength; attenuation constant; charge linear density; desintegration constant; free path; thermal conductivity *Upper case:* equivalent conductivity; permeance; magnetic conductance
M	μ	mu	*Lower case:* amplification factor; permeability; micro- or micron; electric moment; inductivity; magnetic moment; molecular conductivity
N	ν	nu	*Lower case:* reluctivity

Table 1.13 Greek Alphabet (*continued*)

Upper case	Lower case	Name	Common uses
Ξ	ξ	xi	*Lower case:* electromotive force
O	o	omicron	–
Π	π	pi	*Lower case:* 3.141592653589793
			Upper case: product of terms
P	ρ	rho	*Lower case:* resistivity; electric charge density; ripple factor
Σ	σ	sigma	*Lower case:* summation
			Upper case: complex propagation constant; electrical conductivity; leakage constant; standard deviation; surface charge density; Stefan-Boltzmann constant
T	τ	tau	*Upper case:* period
Y	υ	upsilon	–
φ	φ	phi	*Lower case:* angle; phase
			Upper case: magnetic flux; scalar potential; heat current or flow
X	χ	chi	*Lower case:* electric susceptibility
Ψ	ψ	psi	*Lower case:* dielectric flux; angles; electric flux of induction; total flux or electric displacement
W	ω	omega	*Upper case:* resistance in ohms; solid angle
			Lower case: angular speed; angular frequency

INCH/MILLIMETER CONVERSION

One inch is equal to 25.40 millimeters. The following formulas are used to convert millimeters to inches and inches to millimeters.

Millimeters to inches

$$I = 0.038907 \times M \qquad (1.9a)$$

or

$$I = \frac{M}{25.40} \qquad (1.9b)$$

where   M = length in millimeters (mm)
      I = length in inches (in)

Inches to millimeters

$$I = M \times 25.40 \qquad (1.10a)$$

or

$$M = \frac{I}{0.03907} \qquad (1.10b)$$

where the variables are as in the previous equation.

Inches to Millimeters

Table 1.14   Inches to Millimeters

Inches	Millimeters	Inches	Millimeters
0.001	0.0254	0.2	5.080
0.002	0.0508	0.3	7.62
0.003	0.0762	0.4	10.16
0.004	0.1016	0.5	12.70
0.005	0.1270	0.6	15.24
0.006	0.1524	0.7	17.78
0.007	0.1778	0.8	20.32
0.008	0.2032	0.9	22.86
0.009	0.2286	1	25.40
0.01	0.2540	2	50.80
0.02	0.5080	3	76.20
0.03	0.7620	4	101.6
0.04	1.016	5	127.0
0.05	1.270	6	152.4
0.06	1.524	7	177.8
0.07	1.778	8	203.2
0.08	2.032	9	228.6
0.09	2.286	10	254.0
0.1	2.540		

Millimeters to Inches

Table 1.15   Millimeters to Inches

Millimeters	Inches	Millimeters	Inches
0.01	0.000394	0.6	0.02358
0.02	0.000786	0.7	0.02751
0.03	0.001179	0.8	0.03144
0.04	0.001572	0.9	0.03537
0.05	0.001965	1	0.0389
0.06	0.002358	2	0.0786
0.07	0.002751	3	0.1179
0.08	0.003144	4	0.1572
0.09	0.003537	5	0.1965
0.1	0.003937	6	0.2358
0.2	0.00786	7	0.2751
0.3	0.01179	8	0.3144
0.4	0.01572	9	0.3537
0.5	0.01965	10	0.3890

TIME

Important time relationships are shown below.

1 day	=	24 hours = 1440 minutes = 86,400 seconds
1 hour	=	60 minutes = 3600 seconds
1 minute	=	1/60 hours = 60 seconds
1 year	=	31,556,925,975 seconds

PHYSICAL CONSTANTS

Table 1.16 Physical Constants

Constant	Abbrev.	Value	Unit
Speed of light in vacuum	c	2.9979250×10^8	m/s
Elementary charge	e	$1.6021917 \times 10^{-19}$	C
Avogadro constant	N	6.022169×10^{23}	mol^{-1}
Atomic mass unit	u	1.660531×10^{-27}	kg
Electron rest mass	m_e	9.109558×10^{-31}	kg
Proton rest mass	m_p	1.672614×10^{-27}	kg
Faraday constant	F	9.648670×10^4	C/mol
Planck constant	h	6.626196×10^{-34}	J·s
Fine structure constant	α	7.297759	–
Rydeberg constant	$R\infty$	$1.097373\ 12 \times 10^7$	m^{-1}
Bohr magnetron	μB	9.274096×10^{-24}	J/T
Boltzmann's constant	k	1.380622×10^{-23}	J/K
Gravitational constant	G	6.6732×10^{-11}	$N·m^2/kg^2$

TEMPERATURE CONVERSIONS

Degrees Celsius (C), Fahrenheit (F), Kelvin (K), and Reamur (R)

The following formulas are used in temperature conversions. The Celsius scale is also called Centigrade, and the Kelvin scale is also called *absolute*. Figure 1.6 shows the reference points in all scales (boiling and freezing points of water). In the following equations, F = temperature in degrees Fahrenheit, K = temperature in Kelvins,[*] C = temperature in degrees Celsius, and Re = temperature in degrees Reamur.

---

[*]By convention, steps in the Kelvin scale are called "Kelvins" rather than "degrees Kelvin," and the degree symbol is omitted.

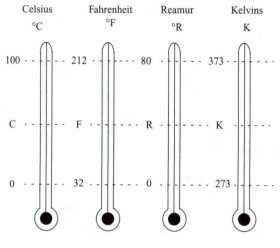

Figure 1.6

Celsius to Fahrenheit

$$F = \left(\frac{9}{5} \times C\right) + 32 \qquad (1.11)$$

Fahrenheit to Celsius

$$C = 5/9 \times (F - 32) \qquad (1.12)$$

Celsius to Kelvin

$$K = C + 273.16 \qquad (1.13)$$

Kelvin to Celsius

$$C = K - 27316 \qquad (1.14)$$

Fahrenheit to Kelvin

$$K = \left[\frac{5}{9}(F - 32)\right](+ 273.16) \qquad (1.15)$$

Kelvins to Fahrenheit

$$F = \left[\frac{9}{5}(K - 273.16)\right] + 32 \qquad (1.16)$$

Celsius to Reamur

$$C = Re \times \frac{5}{4} \qquad (1.17)$$

Reamur to Celsius

$$Re = C \times \frac{4}{5}$$ (1.18)

Fahrenheit to Reamur

$$Re = \frac{4}{9}(F - 32)$$ (1.19)

Reamur to Fahrenheit

$$F = \left(\frac{9}{4} \times Re\right) + 32$$ (1.20)

Kelvins to Reamur

$$Re = \frac{4}{5}(K - 273.16)$$ (1.21)

Reamur to Kelvin

$$K = \left(\frac{5}{4}\right)$$ (1.22)

Celsius to Fahrenheit Conversion Table

Table 1.17   Degrees Celsius to Fahrenheit Conversion

C	F	C	F	C	F	C	F
−50	−58	5	41	55	131	105	221
−45	−49	10	50	60	140	110	230
−40	−40	15	59	65	149	115	239
−35	−31	20	68	70	158	120	248
−30	−22	25	77	75	167	125	257
−25	−13	30	86	80	176	130	266
−20	−4	35	93	85	185	135	275
−15	5	40	104	90	194	140	284
−10	14	45	113	95	203	145	293
−5	23	50	122	100	212	150	302
0	32						

Fahrenheit to Celsius Conversion Table

Table 1.18   Fahrenheit to Celsius Conversion

F	C	F	C	F	C	F	C
−50	−46	35	1.7	95	35	150	66
−40	−40	40	4	100	38	155	68
−30	−34	45	7	105	41	160	71
−20	−29	50	10	110	43	165	74
−10	−23	55	13	115	46	170	77
0	−18	60	16	120	49	175	79
5	−15	65	18	125	52	180	82
10	−12	70	21	130	54	185	85
15	−9	75	24	135	57	190	88
20	−7	80	27	140	60	195	91
25	−4	85	29	145	63	200	93
30	−1	90	32				

Density of Materials

Table 1.19   Density of Materials

Material	Density (g/cm^3)	Material	Density (g/cm^3)
Acetal resin	1.57	Nickel chromium cobalt alloy	8.21
Acrylic	1.19	Nickel iron superalloy	7.86
Aluminium	2.70	Nickel silver	8.70
Aluminium bronze	7.64	Nickel-chromium (Inconel 718)	8.20
Aluminium carbide	2.99	Nickel-moly (Hastelly B-2)	9.20
Aluminium copper alloy	2.84	Niobium nitride	7.30
Aluminium nitride	3.26	Niobium (Columbium)	8.57
Aluminium oxide	3.98	Nylon 6	1.64
Aluminium zinc alloy	2.78	Osmium	22.61
Antimony	6.65	Palladium	12.02
Aramotic polymide	1.44	PE (polyethylene	0.95
Aremid fiber	1.45	Pewter (Sn, Sb, Cu)	7.20
Astempered ductile iron	7.20	Phenolic resin	1.99
Beryllium	1.85	Phosphorus	1.83
Bismuth	9.80	Platinum	21.45
Boron	2.40	Polycarbonate	1.53
Boron carbide	2.52	Polymide thermoset	2.00
Boron nitride	2.25	Polypropylene	1.05

Table 1.19   Density of Materials *(continued)*

Material	Density (g/cm³)	Material	Density (g/cm³)
Brass (61.5Cu-3Pb-35.5Zn)	8.70	Polysulfone	1.24
Bronze (57Cu,40Zn,3Pb)	8.70	Polyurethane	1.27
Cadmium	8.65	Potassium	0.86
Calcium	1.55	Praseodymium	6.77
Carbon fiber	1.74	Pyrex glass	2.52
Cellulose acetate	1.30	Rhenium	21.0
Cerium	6.77	Rhodium	12.41
Cerium dioxide	7.28	Rubidium	1.53
Chromium	7.19	Ruthenium	12.45
Chromium carbide	6.70	Samarium	7.49
Chromium steel	7.83	Silicon	2.33
Cobalt	8.85	Silicon carbide	3.10
Copper	8.96	Silicon nitride	3.19
Copper zinc alloy	8.19	Silicone	1.35
E-glass fiber	2.62	Silver	10.49
Epoxy resin	1.56	Sodium	0.97
Galium	5.91	Stainless steel (17Cr-4Ni)	7.81
Germanium	5.32	Sulfur	2.07
Gold	19.30	Tantalum	16.60
Graphite	2.26	Tantalum cabide (TaC)	14.53
Hafnium	13.10	Thallium	11.85
Hot work tool steel	7.75	Thorium	11.50
Human Bone	1.30	Thulium	9.31
Indium	7.31	Tin	7.30
Iron	7.87	Tin bronze	9.29
Iron-nickel(Invar)	8.00	Titanium	4.51
Lanthanum	6.15	Titanium dioxide	4.25
Lead	11.34	Titanium nitride	5.29
Lithium	0.53	Tungsten	19.40
Magnesium	1.74	Tungsten carbide	17.20
Manganese	7.43	Unsaturated polyester	2.00
Maraing steel	8.02	Uranium	19.07
Molybdenum	10.20	Vanadium	6.11
Neodymium	7.00	Zinc	7.13
Nickel	8.90	Zinconium	6.49
Nickel aluminide (NiAl)	6.05	Zirconium (partially stabilized)	5.70

THERMAL EXPANSION OF SOLIDS

Linear Expansion

A change in the temperature of a solid is accompanied by a change in its linear dimensions and volume. The changes in length depend on the temperature variation and its coefficient of linear expansion (α) and can be calculated using Eq. (1.23).[*]

$$l_t = l_o \times (1 + \alpha t) \tag{1.23}$$

where l_t = the final length of the body
   l_o = the initial length of the body
   t = the temperature change (final minus initial)
   α = the coefficient of linear expansion

Volumetric Expansion

Similarly, a change in the temperature of a solid is accompanied by a change in its volume. The changes will depend on the temperatures (initial and final) and the coefficient of volume expansion (β) and can be calculated using Eq. (1.24).[†]

$$V_t = V_o \times (1 + \beta t) \tag{1.24}$$

where V_t = the final volume of the body
   V_o = the initial volume of the body
   β = the coefficient of volume expansion

Coefficients of Linear Expansion

Table 1.20   Coefficients of Linear Expansion (α) of Solids (at 20°C)[*]

Substance	$\alpha \times 10^{-6}$	Substance	$\alpha \times 10^{-6}$
Aluminium	22.9	Ice (from −10 to 0°C)	50.7
Bismuth	13.4	Iridium	6.5
Brass	18.9	Iron	10.2 to 11.9
Brick Masonry	5.5	Lead	28.3
Bronze	17.5	Magnesium	25.1
Carbon (Graphite)	7.9	Nickel	13.4
Concrete	12.0	Platinum	8.9
Constantan	17.0	Platinum-Iridium Alloy	8.7
Copper	16.7	Porcelain	3.0

[*]The units must be kept consistent; i.e., temperatures in Celsius, lengths in meters.
[†]In this case as well, the units must be kept consistent.

Table 1.20   Coefficients of Linear Expansion (α) of Solids (at 20°C)[*] (continued)

Substance	$\alpha \times 10^{-6}$	Substance	$\alpha \times 10^{-6}$
Diamond	0.91	Quartz (used)	0.5
Duralumin	22.6	Steel	11.0 to 12.0
Ebonite	70	Tin	21.4
Silver	18.4	Tungsten	4.3
Glass (ordinary)	8.5	Vinyl Plastic	70
Glass (pyrex)	3	Wood (accross grain)	50 to 60
Gold	14.5	Wood (along grain)	2 to 6
Granite	8.3	Zinc	30.0

[*] *Obs.:* In some cases, the values in the table depend on the degree of purity of the material.

Coefficients of Volume Expansion

Table 1.21   Coefficients of Volume Expansion (β) of Some Liquids (18°C)

Substance	$\beta \times 10^{-4}$	Substance	$\beta \times 10^{-4}$
Acetone	14.3	Nitric acid	12.4
Aniline	8.5	Petroleum	9.2
Benzene	10.6	Propyl alchool	9.8
Ethyl alchool	11.0	Toluene	10.8
Ethyl ether	16.3	Water (5 to 10°C)	0.53
Glycerine	5.0	Water (10 to 20°C)	1.56
Kerosene	10.0	Water (20 to 40°C)	3.02
Mercury	1.81	Water (40 to 60°C)	4.55
Methyl alcohol	11.9	Water (60 to 80°C)	5.57

BASIC ELECTRICITY

Table 1.22 shows the electrical resistivity of some selected materials.

Table 1.22   Electrical Resistivity of Selected Materials

Material	Resistivity, $\mu\Omega\cdot$cm	Material	Resistivity, $\mu\Omega\cdot$cm
2V Permender	40	Al 2024 (O)	3.4
35Ni-20Cr-45Fe	101.4	Al 5052 (all)	4
35Pd-30Ag-14Cu-10Pt-10Au-1Zn	35	Al 6061 (T6)	4.0
41.7Au-32.5Cu-18.8Ni-7Zn	39	Al 7075 (O)	3.8
45 Permalloy	45	Al7075 (T6)	5.2
45Pd-30Ag-20Au-5Pt	39	Alnico I	75

Table 1.22   Electrical Resistivity of Selected Materials (continued)

Material	Resistivity, µΩ·cm	Material	Resistivity, µΩ·cm
50Pd-50Ag	31.5	Alnico II	65
55Cu-45Ni	49.87	Alnico III	60
55Ni-18Ag	31	Alnico V	47
58Ag-15Cd	4.93	Alnico VI	50
60Ag-40Pd	23	Carbon steel,0.65C	18
60Ni-16Cr-24Fe	112.2	Cu-0.5Te	1.82
60Pd-40Ag	43	Cu-1.0Pb	1.76
60Pd-40Cu	35(c)	Cu-1.2Sn+P	3.6
66.5Fe-22NI-8.5Cr	58.18	Cu-15Zn	4.7
67Cu-5Ni-27Mn	99.74	Cu-1Cr(max)	2.10
67Ni-30Cu-1.4Fe-1Mn	56.52	Cu-20Zn	5.4
69Au-25Ag-6Pt	15	Cu-2Be(max)	11.5
70Ag-30pd	14.3	Cu-5Sn+P	11
70Cu-13Mn (manganin)	48.21	Cunife	18
70Pt-30Ir	35	ETP copper	1.71
71Ni-29Fe	19.95	Fe-17Co	28
72Ag-26Cu-2Ni	2.9	Fe-35Co-1Cr	20
72Fe-23Cr-5Al	138.8	Fe-36Co	27
72Mn-18Cu-10Ni	112.2	Fe-6W	30
75Au-25Ag	10.8	Fe-78Ni	16
75Fe-22Ni-3Cr	15.79	Fine silver	1.59
75Fe-22Ni-3Cr	78.13	Gold	2.35
75Ni-20Cr-3Al	132.98	Graphite	910.1
75Pt-25Ir	33	HH(Fe-26Cr-12Ni)	80
78.5Ni-20Cr-1.5Si(80-20)	108.05	HT(Fe-39Ni-18Cr)	100
78Cu-22Ni	29.92	Iron 99.99%	9.71
80Ni-20Cr	112.2	M-15 steel (max)	69
85Cu-10Mn-4Ni (shunt manganin)	38.23	M-17 steel(max)	58
86Pt-14Ru	46	M-19 steel(max)	56
88Cu-8Sn-4Zn	15	M-27 steel(max)	47
90Ag-10Au	4.2	M-5 steel(max)	52
90Ag-10Cu	2.0	M-50 steel	18
90Ag-10Pd	5.3	Molybdenum	5.2
90Au-10Cu	10.8	Molybdenum disilicide,MoSi2	37.24
90Pd-10Ru	27	Mumetal	60
90Pt-10Ru	43	OF copper	1.71
94Cu-6Ni	9.93	Palladium	10.8
95Pt-5Ir	19	Platinum	10.6
95Pt-5Ru	31.5	Pure copper	1.67

Table 1.22 Electrical Resistivity of Selected Materials (continued)

Material	Resistivity, $\mu\Omega\cdot cm$	Material	Resistivity, $\mu\Omega\cdot cm$
96Pt-4W	36	Silicon carbide(SiC)(max)	200
97Ag-3Pd	2.9	Supermalloy	65
97Ag-3Pt	3.5	Supermendur	40
98Cu-2Ni	4.99	Tantalum	12.45
99.8Ni	7.98	Thermenol	162
AISI 1010 steel	12	Type 309 stainless	78
AISI 1040	56	Yellow brass	6.4
Al 1100(O)	2.9		

DIELECTRIC CONSTANT OF SELECTED MATERIALS

Table 1.23 Dielectric Constant of Selected Materials

Material	Dielectric constant	Material	Dielectric constant	Material	Dielectric constant
ABS pellet	1.5–2.5	Ebonite	2.8–4.5	PE pellet	1.5
Acetone	19.5–20.0	Epoxy resin	2.5–6.0	Plexiglass	3.0–3.5
Acrylic resin	2.7–6.0	Ethyl alcohol	20–27	PP pellet	1.5–1.8
Air	1.0	Ethyl Ether	4.1–4.8	Polystyrene	2.2–2.5
Alcohol, industrial	16–31	Fluor	2.5–3.0	Polyvinyl	3.0–3.6
Aluminum powder	1.6–1.8	Fly ash	1.5–1.7	Porcelain	3.1–6.5
Aluminum sulfate	6	Glass	6–10	Potassium chloride	4.6
Asphalt	2.5–3.2	Glass, raw material	2.0–2.5	PVC powder	1.4
Bakelite	4.5–7.0	Glycerine	50–56	Rice	3–8
Beeswax	2.7–2.9	Grain	3–8	Rosin	2.5–3.5
Benzene, liquid	2.2–2.3	Granulated sugar	1.5–2.2	Rubber, raw	2.1–2.7
Bitumen	2.5–3.3	Heavy oil	2.6–3.0	Sand	3–8
Calcium carbonate	1.8–2.0	Hexane, liquid	5.8–6.3	Silk (natural)	4.5
Calcium oxide	1.8	Hydrochloric acid	4–12	Soap powder	1.2–1.5
Calcium sulfate	5.6	Iron oxide	14.2	Sodium sulfite	5
Carbon dioxide	1.6	Liquid nitrogen	1.4	Starch	2–5
Celluloid	3–4	Kerosene	2.8	Sugar	3
Cement	1.5–2.1	Marble	8–10	Vacuum	0.99
Chlorine water	2	Mica	2.5–8.0	Vegetable oil	2.5–3.5
Coal, powder	1.2–1.8	Mineral oil	2.1	Water	48–80
Coffee, powder	2.4–2.6	Nylon	4–5	Wheat, powder	2.5–3
Coke	1.1–2.2	Paint	5–8	Xylene, liquid	2.2–2
Corn, refuse	2.3–2.6	Parafin	2.0–2.5		

OHM'S LAW

The voltage drop across a resistor is equal the current passing through it times its resistance.

Formula:

$$U = R \times I \qquad (1.25a)$$

$$I = U/R \qquad (1.25b)$$

$$R = U/R \qquad (1.25c)$$

where U = the voltage in volts (V)
R = the resistance in ohms (Ω)
I = the current flowing through the resistor in amperes (A)

JOULE'S LAW

The amount of power consumed by a device or drained from a generator is proportional to the product of the amount of current flowing through the device and the applied voltage.

Formula:

$$P = U \times I \qquad (1.26a)$$

$$P = R \times I^2 \qquad (1.26b)$$

$$P = U^2/R \qquad (1.26c)$$

PARALLEL/SERIES RESISTORS

Figure 1.7 shows resistors wired in series and parallel. The equivalent resistance is calculated by the following formulas.

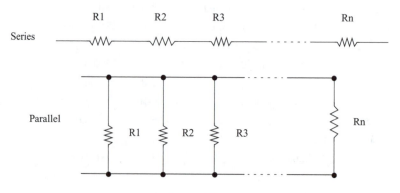

Figure 1.7

(a) Series

$$R = R_1 + R_2 + R3 + \ldots + R_n \tag{1.27}$$

(b) Parallel

$$1/R = 1/R_1 + 1/R_2 + 1/R_3 + \ldots + 1/R_n \tag{1.28}$$

Color Code for Resistors

Resistors are coded with colored bands to indicate their values (Fig. 1.8). The color codes are as shown in Table 1.24.

Table 1.24   Color Code for Resistors

Color	Significant digits	Multiplier	Tolerance (%)	Temperature coefficient (ppm/°C)
Black	0	1	—	—
Brown	1	10	1	100
Red	2	100	2	50
Orange	3	1000	—	15
Yellow	4	10,000	—	25
Green	5	100,000	0.5	—
Blue	6	1,000,000	0.25	10
Violet	7	10,000,000	0.1	5
Gray	8	100,000,000	0.05	—
White	9	1,000,000,000	—	1
Gold	—	0.1	5	—
Silver	—	0.01	10	—

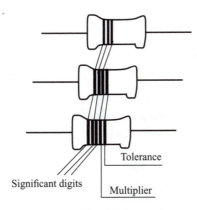

Figure 1.8

CAPACITORS

Capacitance

A planar capacitor is formed by two parallel plates, between which is placed an insulating material called a *dielectric*. Capacitance is calculated using the following formulas.

(a) Capacitance

$$C = 0.08859 \times \varepsilon \times [s(n-1)/d] \tag{1.29}$$

where C = capacitance in picofarads (pF)
   ε = dielectric constant (see Table 1.23)
   s = active surface area of the smaller plate (if the capacitor uses different size plates) in square centimeters (cm^2)
   n = number of plates
   d = distance between plates in centimeters (cm)

(b) Energy stored in a capacitor

$$W = 1/2 \times C \times U^2 \tag{1.30}$$

where W = stored energy in joules (J)
   C = capacitance in farads (F)
   U = voltage between the capacitor's plates in volts (V)

Parallel/Series Capacitors

Capacitors can be wired in series or parallel as shown in Fig. 1.10. The equivalent capacitance for each association is calculated by the following formulas.

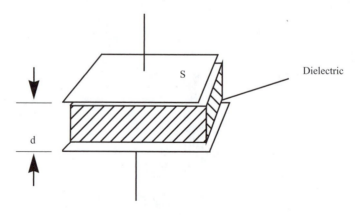

Figure 1.9

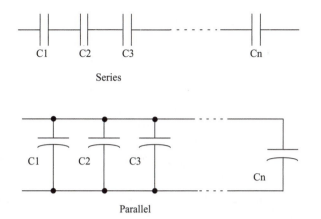

Figure 1.10

(a) Parallel

$$C = C_1 + C_2 + C_3 + \ldots + C_n \tag{1.31}$$

(b) Series

$$1/C = 1/C_1 + 1/C_2 + 1/C_3 + \ldots + 1/C_n \tag{1.32}$$

where C = the equivalent capacitance in pF, nF, or mF
$C_1 \ldots C_n$ = the associated capacitances in the same units

INDUCTANCE

The self-inductance or inductance of an air-core coil can be calculated using the following formula (see Fig. 1.11):

$$L = 1,257 \times (s \times n/m) \times 10^{-8} \tag{1.33}$$

where L = self-induction coefficient in henrys (H)
n = number of turns
S = cross section area of a turn in square centimeters (cm^2)
m = solenoid's length in centimeters (cm)

Parallel/Series Inductances

Coils (inductances) can be wired in series or parallel as shown in Fig. 1.12. The equivalent inductance for each association is calculated using the following formulas.

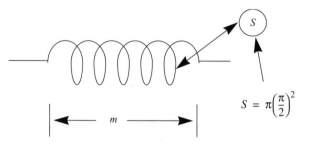

$$S = \pi\left(\frac{\pi}{2}\right)^2$$

Figure 1.11

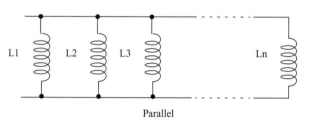

Series

Parallel

Figure 1.12

(a) Series

$$L = L_1 + L_2 + L3 + \dots + L_n \qquad (1.34)$$

(b) Parallel

$$1/L = 1/L_1 + 1/L_2 + 1/L_3 + \dots + 1/L_n \qquad (1.35)$$

where L = equivalent inductance in henrys (H)
          $L_1 \dots L_2$ = associated inductances in henrys (H)

ALTERNATING CURRENT

Formulas:
(a) Frequency and period

$$f = 1/T \qquad (1.36)$$

where f = frequency in hertz (Hz)
        T = period in seconds

(b) Average value for a sinewave (see Fig. 1.13)

$$U_{avg} = 0.637 \times U_{max} \tag{1.37}$$

(c) RMS (root mean square) value for a sinewave (also called effective *value)*

$$U_{rms} = 0,707 \times U_{max} \tag{1.38}$$

where U_{max} = maximum value or peak value in volts (V)
U_{avg} = average value in volts (V)
U_{rms} = root mean square or effective value in volts (V)

(d) Capacitive reactance

$$X_c = 1/(2 \times \pi \times f \times C) \tag{1.39}$$

(e) Inductive reactance

$$X_L = 2 \times \pi \times f \times C \tag{1.40}$$

where X_c = capacitive reactance in ohms (Ω)
X_L = inductive reactance in ohms (Ω)
f = frequency in hertz (Hz)
π = the constant 3.1416
C = capacitance in farads (F)

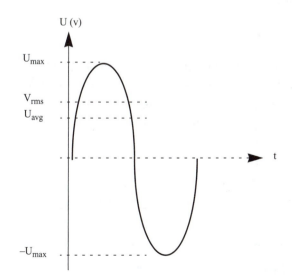

Figure 1.13

(f) Resonance

The frequency at which the capacitive reactance of an inductance-capacitance (LC) circuit becomes equal to the inductive reactance is termed *resonant frequency* and, both for parallel and series circuits, is calculated by the following formula (see Fig. 1.14):

$$fr = \frac{1}{2 \times \pi \times \sqrt{L \times C}} \tag{1.41}$$

where fr = resonant frequency in hertz
π = the constant 3.1416
L = inductance in henrys (H)
C = capacitance in farads (F)

TRANSFORMERS

Formula:
(a) Voltage ratio

$$U_1/U_2 = n_1/n_2 \tag{1.42}$$

where U_1 = voltage applied to the primary winding in volts (V)
U_2 = voltage in the secondary winding in volts (V)
n_1 = number of turns in the primary winding
n_2 = number of turns in the secondary winding

(b) Impedance ratio

$$Z_1/Z_2 = n_1^2/n_2^2 \tag{1.43}$$

where Z_1 = impedance of the primary winding in ohms (Ω)
Z_2 = impedance of the secondary winding in ohms (Ω)
n_1 = number of turns in the primary winding
n_2 = number of turns in the secondary winding

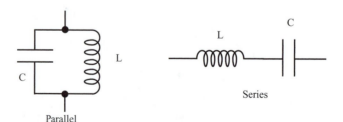

Parallel

Series

Figure 1.14

OTHER TABLES AND INFORMATION

WHEATSTONE BRIDGE

This bridge is formed by resistors and used to perform resistance measurements. The circuit of a Wheatstone bridge is shown in Fig. 1.15. The detecting device can be a galvanometer. The applied signal is a DC voltage.

Obs: When used for resistance measurements, one of the resistances is unknown ($R_4 = R_x$, for instance), and other is variable to balance the bridge (R_3, for instance).

Formula:

$$R_4 = (R_1/R_2) \times R_3 \qquad (1.44)$$

where $R_1 \ldots R_4$ = the resistances in the arms of the bridge in ohms (Ω)

WIRE GAUGE

The following tables are useful when working with enameled wires and solid insulated conductors. They allow the designer to calculate the resistance of a length of wire (such as used in a solenoid, relay, or motor) or to make the correct choice according to the current rating for the application.

WIRE CURRENT RATINGS

Table 1.25   Wire Current Ratings

Minimum wire size (AWG)	Current rating (A)	Minimum wire size (AWG)	Current rating (A)
22	0.5	14	12
20	1	12	20
18	3	10	32
16	6		

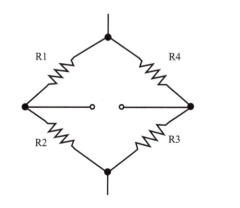

Figure 1.15

STANDARD ANNEALED COPPER WIRE ATTRIBUTES

Table 1.26   Standard Annealed Copper Wire (AWG and B & S)

AWG	Diameter (mm)	Cross-section (mm^2)	Resistance (Ω/km)	AWG	Diameter (mm)	Cross-section (mm^2)	Resistance (Ω/km)
0000	11.86	107.2	0.158	21	0.7230	0.41	41.46
000	10.40	85.3	0.197	22	0.6438	0.33	51.5
00	9.226	67.43	0.252	23	0.5733	0.26	56.4
0	8.252	53.48	0.317	24	0.5106	0.20	85.0
1	7.348	42.41	0.40	25	0.4547	0.16	106.2
2	6.544	33.63	0.50	26	0.4049	0.13	130.7
3	5.827	26.67	0.63	27	0.3606	0.10	170.0
4	5.189	21.15	0.80	28	0.3211	0.08	212.5
5	4.621	16.77	1.01	29	0.2859	0.064	265.6
6	4.115	10.55	1.27	30	0.2546	0.051	333.3
7	3.665	10.55	1.70	31	0.2268	0.040	425.0
8	3.264	8.36	2.03	32	0.2019	0.032	531.2
9	2.906	6.63	2.56	33	0.1798	0.0254	669.3
10	2.588	5.26	3.23	34	0.1601	0.0201	845.8
11	2.305	4.17	4.07	35	0.1426	0.0159	1,069
12	2.053	3.31	5.13	36	0.1270	0.0127	1,339
13	1.828	2.63	6.49	37	0.1131	0.0100	1,700
14	1.628	2.08	8.17	38	0.1007	0.0079	2,152
15	1.450	1.65	10.3	39	0.0897	0.0063	2,669
16	1.291	1.31	12.9	40	0.0799	0.0050	3,400
17	1.150	1.04	16.34	41	0.0711	0.0040	4,250
18	1.024	0.82	20.73	42	0.0633	0.0032	5,312
19	0.9116	0.65	26.15	43	0.0564	0.0025	6,800
20	0.8118	0.52	32.69	44	0.0503	0.0020	8,500

CHEMISTRY

DRYING AGENTS

Table 1.27 Drying Agents

Water remaining after drying[*] in g/m^3 air	Name	Formula
1.4	Copper sulfate (dehydrated)	$CuSO_4$
0.8	Zinc chloride	$ZnCl_2$
0.14–0.25	Calcium chloride	$CaCl_2$
0.16	Sodium hydroxide	$NaOH$
0.008	Magnesium oxide	MgO
0.005	Calcium sulfate (dehydrated)	$CaSO_4$
0.003	Hydrated aluminium	Al_2O_3
0.002	Potassium hydroxide	kOH
0.001	Silica gel	$(SiO_2)x$
0.000025	Phosphorus pentoxide	P_2O_5

[*] At 22°C.

FREEZING MIXTURES

Table 1.28 Freezing Mixtures

Temperature from (°C)	Drop to (°C)	Mixture
+10	−12	$4\ H_2O + 1\ KCl$
+10	−15	$1\ H_2O + 1\ NH_3NO_3$
+8	−24	$1\ H_2O + 1\ NaNO_3 + 1\ NH_4Cl$
0	−21	3 crushed ice + 1 NaCl
0	−39	1.2 crushed ice + $2\ CaCl_2 \cdot 6\ H_2O$
0	−55	1.4 crushed ice + $2\ CaCl_2 \cdot 6\ H_2O$
+15	−78	1 methyl alcohol + 1 solid CO_2

ACID-BASE INDICATORS

Table 1.29 Acid–Base Indicators

Indicator	pH range	Color change
Thymol blue	1.2–2.8	red to yellow
p-dimethyllamino-azobromophenolblue	2.9–4.0	red to orange-yellow
	3.0–4.6	yellow to red-violet

Table 1.29   Acid–Base Indicators (continued)

Indicator	pH range	Color change
congo red	3.0–4.2	blue-violet to red-orange
methyl orange	3.1–4.4	red to yellow-orange
brom cresol green	3.8–5.4	yellow to blue
methyl red	4.4–6.2	red to orange-yellow
litmus	5.0–8.0	red to blue
bromocresol purple	5.2–6.8	yellow to purple
phenol red	6.4–8.2	yellow to red
neutral red	6.4–8.0	blue-red to orange-yellow
cresol red	7.0–8.8	yellow to purple
meta cresol purple	7.4–9.0	yellow to purple
thymol blue	8.0–9.6	yellow to blue
phenolphtalein	8.2–9.8	colorless to red-violet
alizarin yellow 66	10–12.1	light yellow to light-brown

RELATIVE ATOMIC MASSES OF THE CHEMICAL ELEMENTS

Table 1.30   Relative Atomic Masses of the Chemical Elements

Name	Symbol	Atomic number	Atomic mass	Name	Symbol	Atomic number	Atomic mass
Actinium	Ac	89	227.0278	Mercury	Hg	80	200.59*
Aluminum (a)	Al	13	26.98154	Molybdenum	Mo	42	95.94
Americium	Am	95	(243)	Neodymium (c)	Nd	60	144.24*
Antimony	Sb	51	121.75*	Neon (d)	Ne	10	20.179*
Argon (b,c)	Ar	18	39.948*	Neptunium (a)	Np	93	237.0482
Arsenic	As	33	74.9216	Nickel	Ni	28	58.70
Astatine	At	85	(210)	Niobium	Nb	41	92.9064
Barium (c)	Ba	56	137.33	Nitrogen	N	7	14.0067
Berkelium	Bk	97	(247)	Nobelium	No	102	(259)
Beryllium	Be	4	9.01218	Osmium (c)	Os	76	190.2
Bismuth	Bi	83	208.9804	Oxygen (b)	O	8	15.9994*
Boron (b,d)	B	5	10.81	Palladium (b)	Pd	46	106.4
Bromine	Br	35	79.904	Phosphorus	P	15	30.97376
Cadmium (c)	Cd	48	112.41	Platinum	Pt	78	195.09*
Cesium	Cs	55	132.9054	Plutonium	Pu	94	(244)
Calcium (c)	Ca	20	40.08	Polonium	Po	84	(209)
Californium	Cf	98	(251)	Potassium	K	19	39.0983*
Carbon (b)	C	6	12.011	Praseodymium	Pr	59	140.9077

Table 1.30   Relative Atomic Masses of the Chemical Elements *(continued)*

Name	Symbol	Atomic number	Atomic mass	Name	Symbol	Atomic number	Atomic mass
Cerium (c)	Ce	58	140.12	Promethium	Pm	61	(145)
Chlorine	Cl	17	35.453	Protactinium (a)	Pa	91	231.0359
Chromium	Cr	24	51.996	Radium (a,c)	Ra	88	226.0254
Cobalt	Co	27	58.9332	Radon	Rn	86	(222)
Copper (b)	Cu	29	63.546*	Rhenium	Re	75	186.207
Curium	Cm	96	(247)	Rhodium	Rh	45	102.9055
Dysprosium	Dy	66	162.50*	Rubidium (c)	Rb	37	85.4678*
Einsteinium	Es	99	(254)	Ruthenium (c)	Ru	44	101.07*
Erbium	Er	68	167.26*	Samarium (c)	Sm	62	150.4
Europium (c)	Eu	63	151.96	Scandium	Sc	21	44.9559
Fermium	Fm	100	(257)	Selenium	Se	34	78.96*
Fluorine	F	9	18.998403	Silicon	Si	14	28.0855*
Francium	Fr	87	(223)	Silver (c)	Ag	47	107.868
Gadolinium (c)	Gd	64	157.25*	Sodium	Na	11	22.98977
Gallium	Ga	31	69.72	Strontium (c)	Sr	38	87.62
Germanium	Ge	32	72.59*	Sulfur (b)	S	16	32.06
Gold	Au	79	196.9665	Tantalum	Ta	73	180.9479*
Hafnium	Hf	72	178.49*	Technetium	Tc	43	(97)
Helium (c)	He	2	4.00260	Tellurium (c,d)	Te	52	127.60*
Holmium	Ho	67	164.9304	Terbium	Tb	65	158.9254
Hydrogen (b)	H	1	1.0079	Thallium	Tl	81	204.37*
Indium (b)	In	49	114.82	Thorium (a,c)	Th	90	232.0381
Iodine	I	53	126.9045	Thulium	Tm	69	168.9342
Iridium	Ir	77	192.22*	Tin	Sn	50	118.69*
Iron	Fe	26	55.847*	Titanium	Ti	22	47.90*
Krypton (c,d)	Kr	36	83.80	Tungsten	W	74	183.85*
Lanthanum (c)	La	57	138.9055*	Uranium (c,d)	U	92	238.029
Lawrencium	Lr	103	(260)	Vanadium	V	23	50.9414*
Lead (b,c)	Pb	82	207.2	Xenon (c,d)	Xe	54	131.30
Lithium (b,c,d)	Li	3	6.941*	Ytterbium	Yb	70	173.04*
Lutetium	Lu	71	174.97	Yttrium	Y	39	88.9059
Magnesium (c)	Mg	12	24.305	Zinc	Zn	30	65.38
Manganese	Mn	25	54.9380	Zirconium (c)	Zr	40	91.22
Mendelevium	Md	101	(258)				

CHAPTER 2

Basic Mechanics

In this chapter, you will find formulas and useful information about calculations that are applicable to simple mechanical machines such as levers, pulleys, springs, grippers, and so on. Although other common units are indicated, we recommend adoption of SI units (see Chapter 1 for details).

LEVERS

There are three classes of levers, designated by the location of the *fulcrum* (the pivot or fixed point) in relation to the *resistance* (or *weight*) and the *effort*. Figure 2.1 shows the three classes of levers.

Properties

The ratio of the effort arm length to the resistance arm length is inversely proportional to the ratio of the resistance to the effort required to overcome it. For example, if the effort arm is twice the length of the resistance arm, the effort will be half of the resistance. Figure 2.2 shows a practical example of a first-class lever.

Formula: (see Fig. 2.1)

$$L/\ell = R/E \tag{2.1}$$

where L = length of effort arm
         ℓ = length of resistance arm
         R = resistance weight or force
         E = effort force

MECHANICAL ADVANTAGE

The ratio of the required effort to the weight of the object to be moved gives the mechanical advantage (MA) of a lever.

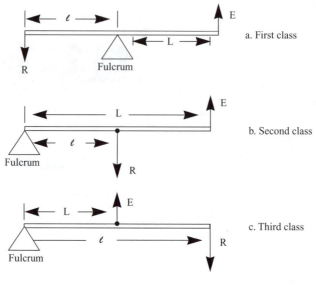

Figure 2.1

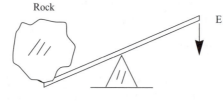

Figure 2.2

Formula: (see Fig. 2.3)

$$MA = R/E \qquad\qquad (2.2)$$

where MA = mechanical advantage
$\quad\quad R$ = resistance weight or force
$\quad\quad E$ = effort force

If $MA > 1$, the lever has a positive gain of force. If $MA < 1$, the lever has a negative gain of force.

Properties

1. Second-class levers have a positive mechanical advantage, since the they can be used to magnify the applied force.

2. Third-class levers have negative mechanical advantage, since you need to apply effort that exceeds the weight to be moved.

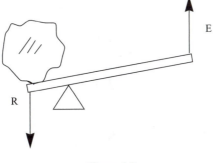

Figure 2.3

3. First-class levers can have positive or negative mechanical advantage, depending on the length of the arms.

GRIPPERS

Grippers may be considered as levers and, depending on the gripper's geometry, one of the previous formulas applies. Figure 2.4 shows the main types of grippers. Equation (2.3) is the formula used to calculate the forces and lengths.

Formula:

$$F_1/F_2 = a/b \tag{2.3}$$

where F_1 and F_2 = forces involved
a and b = lengths of the arms

THE BLOCK AND TACKLE

The block and tackle is a simple mechanical machine that uses the parts shown in Fig. 2.5. Figure 2.6 shows how the block and tackle is used.

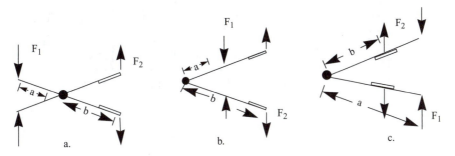

Figure 2.4

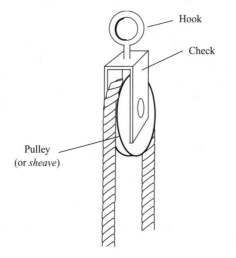

Hook

Check

Pulley
(or *sheave*)

Figure 2.5

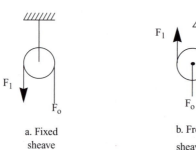

F_1

F_o

a. Fixed
sheave

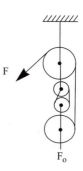

F_1

F_o

b. Free
sheave

F

F_o

c. Ordinary

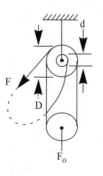

d

F

D

F_o

d. Differential

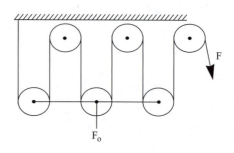

F

F_o

e. Multiple differential

Figure 2.6

The mechanical advantage depends on the type of device, as shown below.

FORCE[*]

Formulas:
Fixed sheave or pulley

$$F_1 = F_o \tag{2.4}$$

Free sheave or pulley

$$F_1 = F_o/2 \tag{2.5}$$

Ordinary pulley block

$$F_1 - F_o/n \tag{2.6}$$

Differential pulley block

$$F_1 = 1/2(1 - d/D)F_o \tag{2.7}$$

Multiple differential pulley block

$$F_1 = F_o/2n \tag{2.8}$$

where F_1 = force required to raise or lower the load
  F_o = load
  n = number of pulley
  d, D = diameters of the pulleys

PATH OF THE LOAD

Formulas: (see Fig. 2.7)
Fixed sheave

$$s = h \tag{2.9}$$

Free sheave

$$s = 2 \times h \tag{2.10}$$

Ordinary pulley block

$$s = n \times h \tag{2.11}$$

*These calculations disregard bearing friction and rope rigidity.

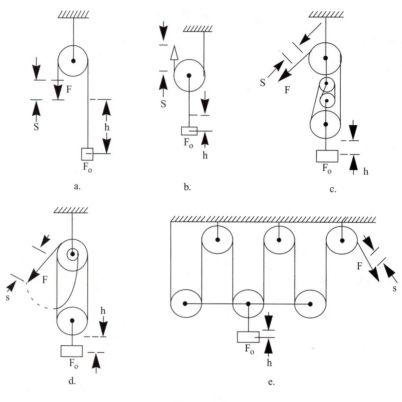

Figure 2.7

Differential pulley block

$$s = 2h/(1 + (-d/D)) \qquad (2.12)$$

Multiple differential pulley block

$$h = D/2n \qquad (2.13)$$

MOMENT OF FORCE

The moment M of a force F about a point O is the perpendicular distance from point O to the line of action of force F times the force F (see Fig. 2.8).

Formula:

$$M = F \times d \qquad (2.14)$$

where M = moment
F = force
d = distance

The resultant forces are shown in Fig. 2.9.

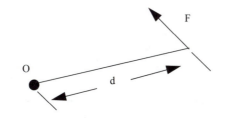

Figure 2.8

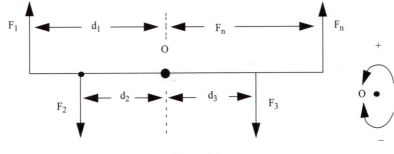

Figure 2.9

Formula:

$$F_1 \times d_1 + F_2 \times d_2 + F_3 \times d_3 + \dots F_n \times d_n = 0 \qquad (2.15)$$

Obs.: F may be positive or negative according to the direction—clockwise or counterclockwise.

THE INCLINED PLANE AND THE WEDGE

The inclined plane permits us to overcome a large resistance by applying a small force over a longer distance to raise a load (see Fig. 2.10).

Formula:

$$F = (R \times h)/L \qquad (2.16)$$

where L = length of the ramp
h = height of the ramp
R = weight of the object to be raised or lowered
F = force required to raise or lower the object

THE WEDGE

The wedge can be considered as a special application of the inclined plane (see Fig. 2.11).

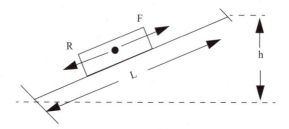

Figure 2.10

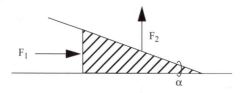

Figure 2.11

Formula:

$$F_1 = F_2 \times \tan \alpha \qquad (2.17)$$

where F_1, F_2, and F = involved forces, in newtons

α = wedge angle

The same formula can be applied to calculate the resultant force R of a wedge.

Obs.: Friction is not considered.

THE SCREW

The screw can be considered as a modified inclined plane. You can make an analogy with the inclined plane by cutting a piece of paper in the shape of a triangle and winding it around a pencil as shown in Fig. 2.12.

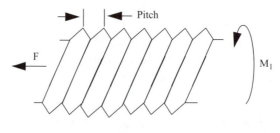

Figure 2.12

Formula:

$$M_1 = F \times r \times \tan(\alpha - \sigma) \tag{2.18}$$

where M_1 = turning moment
F = applied force
r = mean radius of thread
σ = angle of friction ($\tan \sigma = \mu$, where μ is the friction coefficient)

THE JACKSCREW

The jackscrew is used to raise heavy loads by taking advantage of the theoretical mechanical advantage (TMA) of the screw (see Fig. 2.13).

Formula:
(a) Theoretical mechanical advantage

$$\text{TMA} = 1\pi r / \rho \tag{2.19}$$

where TMA = theoretical mechanical advantage
r = length of the handle (m)
ρ = pitch, or distance between corresponding points or successive threads (m)

(b) Force

$$F = F_o h / 2R \tag{2.20}$$

where F = force applied to the circumference of the screw
F_o = resultant force
R = length of the arm of applied force
h = pitch of the screw

Obs.: Here, we assume the absence of friction.

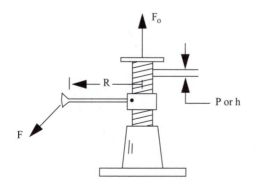

Figure 2.13

GEARS

There are two basic gear arrangements, as shown in Fig. 2.14.

Obs.: When two gears of unequal size are combined, the smaller one is usually called a *pinion*. Figure 2.15 shows other types of gears. Shown are (a) helical spur, (b) rack and pinion, (c) sector, (d) linear, (e) miter, and (f) bevel gear arrangements.

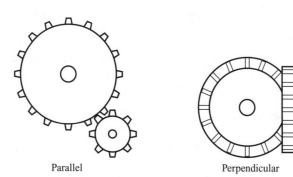

Parallel   Perpendicular

Figure 2.14

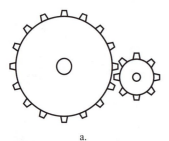

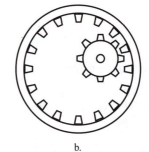

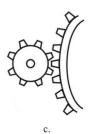

a.   b.   c.

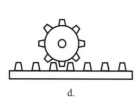

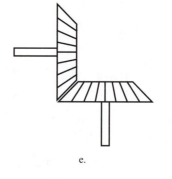

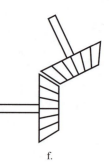

d.   e.   f.

Figure 2.15

Formulas:
(a) Speed change

$$V_1/V_2 = n_1/n_2 \ \text{(linear)}$$

$$\omega_2/\omega_1 = n_2/n_1 \ \text{(angular)} \tag{2.21}$$

where　V_1 and V_2 = tangential speed of the gears
　　　n_1 and n_2 = number of teeth
　　　ω_1, ω_2 = tangential speed in rad/sec

(b) Mechanical advantage

$$\text{TMA} = V_2/V_1 \tag{2.22a}$$

$$\text{TMA} = n_2/n_1 \tag{2.22b}$$

or

$$\text{TMA} = \omega_2/\omega_1 \tag{2.22c}$$

where　　　TMA = theoretical mechanical advantage
　　　V_1 and V_2 = tangential speed of the gears
　　　n_1 and n_2 = number of teeth
　　　ω_1, ω_2 = tangential speed in rad/sec

(c) Torque change

$$\text{TMA} = n_1/n_2 \tag{2.23}$$

where　n_1 = number of teeth on the driven gear
　　　n_2 = number of teeth on the driven gear

$$M_2/M_1 = n_2/n_1 \tag{2.24}$$

$$M_2/M_1 = V_1/V_2 \tag{2.25}$$

$$M_2/M_1 = \omega_1/\omega_2 \tag{2.26}$$

where　M_1 and M_2 = turning moment
　　　V_1 and V_2 = tangential speed of the gears
　　　n_1 and n_2 = number of teeth
　　　ω_1, ω_2 = tangential speed

Obs.: Friction is not considered.

(d) Teeth versus diameter

$$n_1/n_2 = d/D \qquad\qquad (2.27)$$

where n_1 and n_2 = number of teeth

  d and D = diameter of the gears

Figure 2.16 shows all the variables used in the previous formulas. Table 2.1 shows various speed equivalents.

Table 2.1   Speed Unit Equivalence

Units	m/s	m/min	km/h	ft/s	ft/min	mile/h	knot
m/s	1	60	3.60	3.281	196.85	2.237	1.943
m/min	0.01667	1	0.050	0.05468	3.281	0.03728	0.03238
km/h	0.02778	16.67	1	0.9113	54.68	0.6214	0.5396
ft/s	0.3048	18.288	1097	1	60	0.6818	0.5921
ft/min	0.0051	0.3048	0.01687	0.0183	1	0.01136	0.00987
mile/h	0.4470	26.42	1.609	1.467	88	1	0.8684
knot	0.5148	30.887	1.8532	1.6889	101.333	1.1515	1

GEARBOXES

Gearboxes are used to change speed or torque in mechatronic projects. The number and size of the gears and the number of teeth determine the TMA of a gearbox, or the number of times the torque or the speed is multiplied. Figure 2.17 shows some gearboxes used to change speed and increase mechanical advantage.

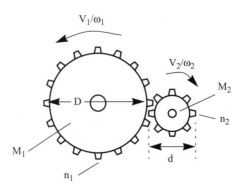

Figure 2.16

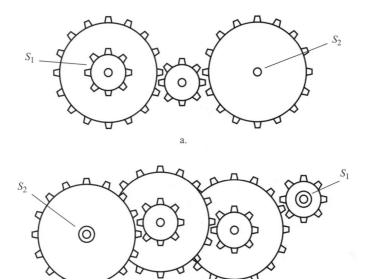

a.

b.

Figure 2.17

Formulas:
(a) Changing the speed (see Fig. 2.18)

$$S_2 = S_1 \times T_1 / T_2 \qquad (2.28)$$

where S_1 = speed of the first shaft in the gear train
S_2 = speed of the last shaft in the gear train

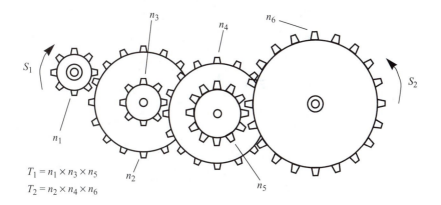

$T_1 = n_1 \times n_3 \times n_5$
$T_2 = n_2 \times n_4 \times n_6$

Figure 2.18

T_1 = product of the teeth on all drivers

T_2 = product of teeth on all driven gears

Speed can be calculated in revolutions per second, cm/s (tangential), or degrees per second (angular).

(b) Changing the torque (see Fig. 2.19)

$$\text{TMA} = T_1/T_2 \tag{2.29}$$

where T_1 = the product of the teeth on all drivers

T_2 = the product of teeth in all driven gears

TMA = the theoretical mechanical advantage

Obs.: Friction is not considered.

THE BEVEL GEAR

Bevel gears are used if the shafts are not in parallel, as shown in Fig. 2.20. They are used to change the direction of a motion. Figure 2.20 shows two main types of bevel gears: (a) the miter gear and (b) the spiral or worm gear.

For the miter gear, the formulas given previously are valid. For the spiral gear, the following formula applies.

Formula:

$$V = n_1 \times 2 \times \pi \times R \tag{2.30}$$

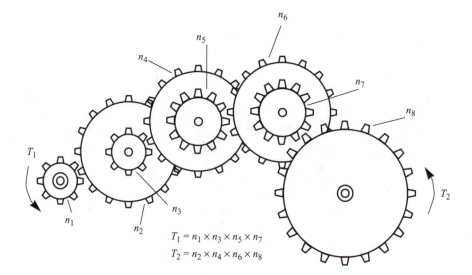

$T_1 = n_1 \times n_3 \times n_5 \times n_7$

$T_2 = n_2 \times n_4 \times n_6 \times n_8$

Figure 2.19

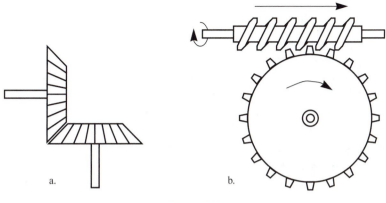

Figure 2.20

where V = the tangential speed of the circular gear

         n_1 = the number of steps advanced by the worm in one turn or revolution

         π = the constant 3.14

         R = the radius of the circular gear (diameter/2)

PULLEYS

Pulleys are used to change speed or torque when used with motors or other rotary power sources. The following formulas are used to calculate the theoretical mechanical advantage of pulley systems and the change of speed. Figure 2.21 shows (a) belt and pulley, (b) O-ring, and (c) chain drive arrangements.

Formulas:

(a) Change of speed

Tangential

$$v_1/v_2 = d_2/d_1 \qquad (2.31)$$

Angular

$$\omega_2/\omega_2 = d_2/d_1 \qquad (2.32)$$

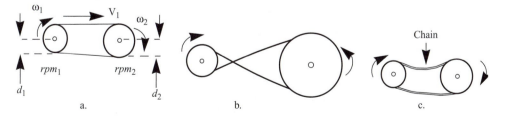

Figure 2.21

rpm

$$rpm_1/rpm_2 = d_2/d_1 \tag{2.33}$$

where v_1 and v_2 = tangential speed of the pulleys
ω_1 and ω_2 = angular speed of the pulleys
rpm_1 and rpm_2 = the revolution per minute of the pulleys
d_1 and d_2 = the diameters of the pulleys

Obs.: The speed of the belt is constant.

(b) Change of torque

$$M_1/M_2 = d_2/d_2 \tag{2.34}$$

where M_1 and M_2 = torque
d_1 and d_2 = diameters of the pulleys

(c) TMA

$$TMA = M_2/M_1 = d_1/d_2 = \omega_1/\omega_2 = rpm_2/rpm_2 = v_2/v_1 \tag{2.35}$$

where   TMA = theoretical mechanical advantage
v_1 and v_2 = tangential speed of the pulleys
ω_1 and ω_2 = angular speed of the pulleys
rpm_1 and rpm_2 = revolution per minute of the pulleys
d_1 and d_2 = diameters of the pulleys

(d) Complex systems

Figure 2.22 shows a complex system using pulleys and belts. The following formulas are valid for calculus involving TMA, change of speed, and torque.

Formulas:
(a) Change of speed

Tangential/angular

$$S_2 = S_1 \times nD/nd \tag{2.36}$$

where S_1 = speed of the first pulley in train
S_2 = speed of the last pulley in train
nD = product of the diameters on all drivers
nd = product of the diameters on all driven gears

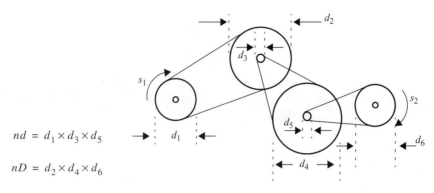

$$nd = d_1 \times d_3 \times d_5$$

$$nD = d_2 \times d_4 \times d_6$$

Figure 2.22

(b) Change of torque

$$M_2/M_2 = S_1/S_2 \qquad (2.37)$$

where M_1 and M_2 = torque of the pulleys

S_1 = speed of the first pulley in the train (angular or tangential)

S_2 = speed of the last pulley in the train (angular or tangential)

(c) TMA

$$\text{TMA} = M_2/M_2 = S_2/S_1 \qquad (2.38)$$

MEASURING THE TORQUE AND SPEED OF A GEAR OR WHEEL

In Chapter 4, the reader will find some practical ways to measure the speed and torque of electric motors (DC and stepper). The same procedures are valid for pulleys, gears, wheels, and any other rotary mechanical device.

SPRINGS

Springs are elastic devices that can be twisted, stretched, or pulled by external forces. They return to their original shape when the external force is released. Figure 2.23 shows several types of springs: (a) flat, (b) spiral, (c) helical, and (d) torsion.

Properties

1. Springs can store energy.

2. Common types of springs are made of steel, phosphor bronze, brass, and other alloys.

a.   b.   c.   d.

Figure 2.23

Formulas:
(a) Spring rate—constant (see Fig. 2.24)

$$R = F/S \tag{2.39}$$

where R = spring rate
F = applied force
S = deformation

Obs.: The general characteristic is given by $R = dF/ds$.

(b) Spring work—constant (see Fig. 2.25)

$$W = F \times (S/2) \tag{2.40}$$

where W = work in joules
s = deformation in meters
F = force in newtons

Obs.: The general characteristic is given by $W = \int F \cdot ds$.

SPRINGS IN SERIES AND PARALLEL
As shown in Fig. 2.26, springs can be used in series and parallel (analogous to resistors and capacitors).

Formulas:
(a) Springs in series

$$S_{tot} = s_1 + s_2 + s_3 + \ldots + s_n \tag{2.41a}$$

Figure 2.24   **Figure 2.25**

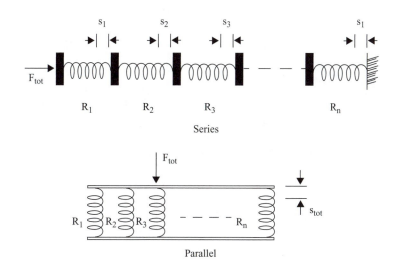

Series

Parallel

Figure 2.26

$$F_{tot} = F_1 = F_2 = F_3 = ... = F_n \qquad (2.41b)$$

$$R_{tot} = R_1 + R_2 + R_3 + ... + R_n \qquad (2.41c)$$

where $S_{tot}, s_1, s_2, s_3, s_n$ = the deformations
$F_{tot}, F_1, F_2, F_3, F_n$ = the involved forces
$R_{tot}, R_1, R_2, R_3, R_n$ = the spring rates

(b) Springs in parallel

$$S_{tot} = S_1 = S_2 = S_3 = ... = S_n \qquad (2.42a)$$

$$F_{tot} = F_1 + F_2 + F_3 + ... + F_n \qquad (2.42b)$$

$$1/R_{tot} = 1/R_1 + 1/R_2 + 1/R_3 + ... + 1/R_n \qquad (2.42c)$$

Obs.: The formulas are valid for compression and tension.

HOW TO DETERMINE THE CHARACTERISTICS OF A SPRING

Figure 2.27 shows simple processes to determine the characteristics of a spring: (a) using a dynamometer and (b) using known weights.

WORK

Work is achieved when a resistance is overcome by a force acting over a measurable distance.

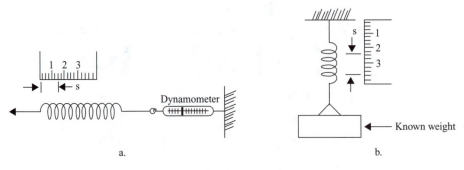

a.　　　　　　　　　　　　　　　　b.

Figure 2.27

Formula:

$$W = F \times d \tag{2.43}$$

where　W = work in joules
　　　　F = force in newtons
　　　　d = distance in meters

Note: The force can be measured in pounds, the work in pound-inches, and the distance in inches.

We can calculate the work of forces in a lever (see Fig. 2.28) as shown below.

Formula:

$$F_1 \times d_1 = F_2 \times d_2 \tag{2.44}$$

where　F_1 = effort applied
　　　　F_2 = resistance overcome
　　　　d_1 = distance over which the effort moves
　　　　d_2 = distance the resistance is moved

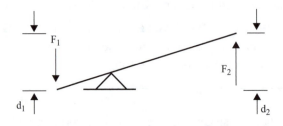

Figure 2.28

FRICTION

Friction is the resistance that one surface offers against its movement over another surface. The amount of friction depends on the nature of the surfaces and the forces that hold them together (see Fig. 2.29).

COEFFICIENT OF SLIDING FRICTION

The following table shows the coefficients of sliding friction for several common materials.

Table 2.2   Coefficient of Sliding Friction for Selected Materials

Surface in contact	Coefficient of friction	Surface in contact	Coefficient of friction
Bronze on bronze	0.2	Leather belt on oak	0.27 to 0.38
Bronze on steel	0.18	Oak on oak	0.5 to 0.6
Cast iron on bronze	0.21	Rubber (tires) on cast iron	0.83
Cast iron on cast iron	0.16	Rubber (tires) on hard soil	0.4 to 0.6
Copper on cast iron	0.27	Sliding bearing greased	0.02 to 0.08
Dry wood on wood	0.25 to 0.5	Steel on iron	0.16
Greased leather belt on metal	0.23	Wooden runners on snow and ice	0.035
Ice on ice	0.028		

POWER

Power can be defined as the work released per unit of time.

Formula:

$$P = W/t \qquad (2.45)$$

where W = work
P = power
t = time

POWER UNIT CONVERSION

Table 2.3 provides conversions for various power units.

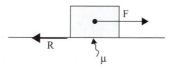

Figure 2.29

Table 2.3   Power Unit Conversion

Units*	hp	BTU/s	ft·lb/s	kW	Poncelet	Cal/s	kgm/s
hp	1	0.7067	550	0.74570	0.76040	0.1781	76.041
BTU/s	1.415	1	778.2	1.055	1.076	0.2520	107.60
ft·lb/s	0.018182	0.001285	1	0.001358	0.00138	0.000323	0.13825
kW	1.34102	0.9480	737.562	1	1.0197	0.2389	101.9716
Poncelet	1.3151	0.9294	723.30	0.9806	1	0.2342	100.0
cal/s	5.615	0.009294	7.3756	0.009807	0.01	0.002342	1

* hp = horsepower, BTU = British thermal unit, ft·lb/s = foot-pounds/second, kW = kilowatt, cal/s = calories per second, kgm/s = kilogram-meters per second.

HORSEPOWER (HP) TO WATT CONVERSION

Considering that 1 hp corresponds to 0.7457 kW or 745.7 W,

Formula:

$$P_{hp} = P_W/745.7 \tag{2.46a}$$

or

$$P_W = P_{hp} \times 745.7 \tag{2.46b}$$

where P_{hp} = power in horsepower (hp)
P_W = power in watts (W)

Various watt/horsepower equivalents are shown in Table 2.4.

Table 2.4   Watt/Horsepower Equivalents

W	hp	W	hp	W	hp
1	0.001341	30	0.040230	500	0.67050
2	0.002682	40	0.053640	600	0.08460
3	0.004023	50	0.067050	700	0.93870
4	0.005364	60	0.080460	800	1.07280
5	0.006705	70	0.093870	900	1.20690
6	0.008046	80	0.107280	1000	1.3410
7	0.009387	90	0.120690	2000	2.6820
8	0.010728	100	0.13410	3000	4.0230
9	0.012069	200	0.26820	4000	5.3640
10	0.013410	300	0.40230	5000	6.7050
20	0.026820	400	0.53640		

OTHER MECHANICAL DEVICES

Gears, pulleys, and chains can be used to change the speed of a shaft and to transfer motion. The following are some techniques used to accomplish this task.

MOTION CONVERSION

(a) Rotary to longitudinal motion using a shaft and a piston

Figure 2.30 shows how a crankshaft and a piston can be used to convert rotary motion into longitudinal motion.

Formula:

$$D = 2 \times R \tag{2.47}$$

where D = stroke of the piston
R = length of the crankshaft

(b) Cams used for motion conversion

Figure 2.31 shows how a rotating cam can be used to convert rotary motion into longitudinal motion.

Formula:

$$D = R - r \tag{2.48}$$

where D = displacement of the pin
R and r = radii of the cam

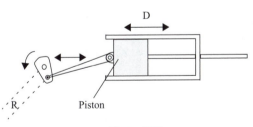

Figure 2.30

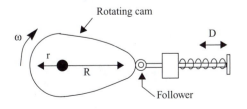

Figure 2.31

(c) Cams used for complex rotary to longitudinal motion conversion

Complex longitudinal motion can be obtained using rotating cams with programmed formats such as the one shown in Fig. 2.32. The movement can be described by tracing the shape of the cam.

(d) Rotary to linear motion conversion using a swash plate

A surface follower contacting a swash plane can be used to convert rotary into linear motion as shown in Fig. 2.33.

(e) Acme screw and rider nut

This is an interesting technique to convert rotary motion into longitudinal motion (see Fig. 2.34).

Formula:

$$D = n \times 2 \times \pi \times d \tag{2.49}$$

where D = total displacement of the nut
n = number of turns of the rotating screw
d = pitch of the screw

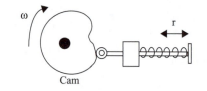

Figure 2.32

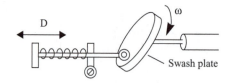

Figure 2.33

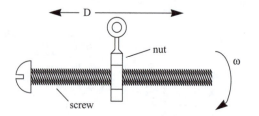

Figure 2.34

(f) Rack and pinion

The combined rack and pinon can be used to convert rotary movement into horizontal, vertical, or inclined movement, and vice-versa. Figure 2.35 shows how the rack and pinion are used.

Formula:

$$V = \omega \times R \qquad (2.50)$$

where V = tangential speed
   ω = angular speed
   R = radius of the pinion

Obs.: $\omega = 2 \times \pi \times f.$

where $\pi = 3.14$
   ω = angular speed
   f = frequency (rotations per second)

ONE-WAY MOTION CONTROLS

One-way motion controls are used to protect components, to limit power dissipation, and to provide a margin of safety. Figure 2.36 shows some ways to implement one-way motion in mechatronics projects: (a) toothed wheel and pawl, and (b) braking pawl.

SHAFT COUPLING

Figure 2.37 gives examples of shaft coupling techniques: (a) rigid coupler, (b) flexible coupler, (c) flexible coupler using rubber tubing, (d) magnetic coupler, and (e) spring coupler.

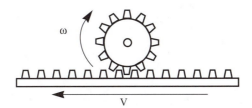

Figure 2.35

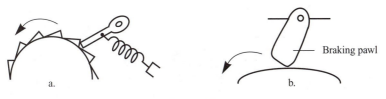

Figure 2.36

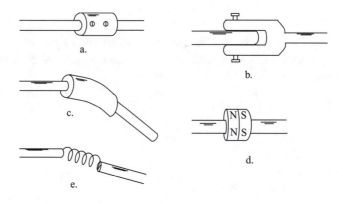

Figure 2.37

CHAPTER 3

Hydrotronics/Pneutronics

HYDROTRONICS

MECHANICS OF LIQUIDS

When using hydraulics or hydrotronics[*] in a mechatronics project, the designer must understand how the devices work, and this implies knowledge of the fundamental concepts and laws relating to the mechanical behavior of liquids. In this chapter, we will provide some formulas, tables, applications, and basic concepts involving hydrotronics.

Pressure

(a) Pressure (i.e., the amount of force per unit of area)

Formula:

$$P = F/S \qquad\qquad (3.1)$$

where P = pressure
F = force
S = area where the force is applied

(b) Pressure of a column of liquid

Formula:

$$P = \rho \times g \times h \qquad\qquad (3.2)$$

[*]We define *hydrotronics* as a science that is derived from the simultaneous application of electronics and hydraulics.

where P = pressure

ρ = density of the liquid

g = acceleration due to gravity (9.8 m/s^2)

h = height of the column

(c) Pressure at a specified depth in a liquid

Formula:

$$P = h \times D \tag{3.3}$$

where P = pressure

h = depth of the point under consideration

D = density of the water (or other liquid)

Water Density versus Temperature

Table 3.1   Density of Water at Various Temperatures

T (°C)	D (g/mL)	T (°C)	D (g/mL)	T (°C)	D (g/mL)
0	0.99987	11	0.99963	21	0.99702
1	0.99993	12	0.99952	22	0.99780
2	0.99997	13	0.99940	23	0.99757
3	0.99999	14	0.00927	24	0.99732
4	1.00000	15	0.99913	25	0.99707
5	0.99999	16	0.99897	26	0.99681
6	0.99997	17	0.99880	27	0.99652
7	0.99993	18	0.99862	28	0.99622
8	0.99988	19	0.99843	29	0.99592
9	0.99981	20	0.99823	30	0.99561
10	0.99973				

Density of Various Liquids

Table 3.2   Density of Various Liquids (at 20°C)

Liquid	Density (g/cm^3)	Liquid	Density (g/cm^3)
Acetic acid	1.049	Heptane	0.684
Acetone	0.791	Hexane	0.660
Aniline	1.02	Machine oil	0.8–0.9
Benzene	0.879	Mercury	13.55
Benzine	0.68–0.72	Methyl alcohol	0.792
Chloroform	1.489	Milk (average)	1.03

Table 3.2 Density of Various Liquids (at 20°C) *(continued)*

Liquid	Density (g/cm³)	Liquid	Density (g/cm³)
Crude oil	0.76–0.86	Nitric acid	1.51
Ethyl alcohol	0.79	Nitroglycerine	1.6
Formic acid	1.22	Sea water	1.01–1.03
Glycerine	1.26	Water	0.99823
Heavy water	1.1086		

Archimedes' Principle

A body immersed in a liquid is buoyed by a force equal to the weight of the liquid that it displaces (see Fig. 3.1).

Communicating Vessels

The heights of the columns in connected vessels are inversely proportional to their densities (see Fig. 3.2).

Formula:

$$h_1/h_2 = d_2/d_1 \qquad (3.4)$$

where h_1 and h_2 = heights of columns
d_1, d_2 = densities of the liquids

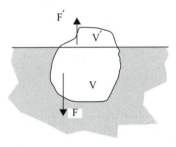

Figure 3.1

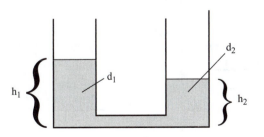

Figure 3.2

FLOW OF LIQUIDS FROM CONTAINERS

The speed of a liquid as it passes through an outlet depends on several factors, including the container shape, liquid density, position of the outlet in the container, and so on. Figure 3.3 shows some cases associated with the formulas below.

(a) Base aperture

Formula:

$$v = Cv \cdot A \cdot \sqrt{2 \cdot g \cdot h} \tag{3.5}$$

(b) Small lateral aperture

Formula:

$$v = Cv \cdot \sqrt{2 \cdot g \cdot h} \tag{3.6}$$

(c) Excess pressure on surface of liquid

Formula:

$$v = Cv \cdot \sqrt{2\left(g \cdot h + \frac{P_{ex}}{\rho}\right)}$$

(d) Excess of pressure applied to an outlet point

Formula:

$$v = Cv \cdot \sqrt{2 \cdot \frac{P_{ex}}{\rho}}$$

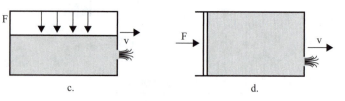

Figure 3.3

where v = velocity in the outlet

Cv = velocity coefficient (0.97 for water)

g = acceleration due to gravity (9.8 m/s^2)

h = height of the container filled with water

P_{ex}/ρ = excess pressure by area unit

USING HYDRAULIC PRESSURE

Hydraulic machines can be used to obtain a mechanical advantage. In practice, we use pistons, but we can basically understand how they work by looking at simple tubes as shown in Fig. 3.4.

Mechanical advantage can be obtained using the configuration shown in the figure, and it can be calculated using the following formulas.

Formulas:

(a) Force gain

$$F_1/F_2 = S_1/S_2 \tag{3.7}$$

where F_1 = force applied to the small piston

F_2 = force applied to the large piston

S_1 = area of the small piston

S_2 = area of the large piston

(b) Displacement

$$F_1/F_2 = d_2/d_1 \tag{3.8}$$

where F_1 = force applied to the small piston

F_2 = force applied to the large piston

d_1 = displacement of the small piston

d_2 = displacement of the large piston

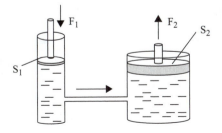

Figure 3.4

(c) Theoretical mechanical advantage (TMA)

$$TMA = F_2/F_2 = d_1/d_2 = S_2/S_1 \tag{3.9}$$

where F_1 = force applied to the small piston
F_2 = force applied to the large piston
d_1 = displacement of the small piston
d_2 = displacement of the large piston
S_1 = area of the small piston
S_2 = area of the large piston

VALVES

Valves are used to limit the flow of a liquid to one direction. Figure 3.6 shows the typical construction of some typical valves: (a) automatic (self-actuated), (b) manual, (c) shunt, and (d) solenoid.

PRACTICAL HYDRAULIC DEVICES

Using solenoids combined with pistons and valves, we can create devices that have applications in mechatronics. Figure 3.6 shows some of these combinations that have practical design applications in: (a) solenoid valve, (b) solenoid actuator, (c) hydraulic switch, (d) electric pump, (e) rotary-to-linear hydraulic converter, and (f) hydraulic motor.

See Chapters 5 and 6 for practical electronic circuits used to drive solenoids.

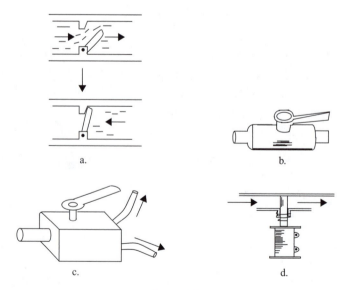

a.

b.

c.

d.

Figure 3.5

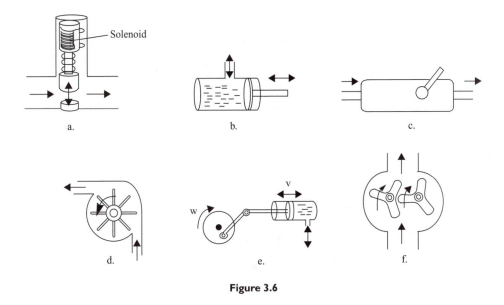

Figure 3.6

PNEUTRONICS

MECHANICS OF GASES

When using pneumatics in mechatronic projects, the designer must understand how the devices work, and this implies a knowledge of the fundamental concepts and laws governing the mechanical behavior of gases. In the following text, we provide some formulas, tables, applications, and basic concepts involving pneumatics and pneutronics.[*]

Pressure

Pressure refers to the amount of force per unit of area.

Formula:

$$P = F/S \tag{3.10}$$

where P = pressure
F = force
S = area

Air Pistons

Pistons can be used to change forces or displacements as shown in Fig. 3.7. Tubes with pistons can also be used to transmit forces from one place to another as shown in the same figure.

---

[*]We define *pneutronics* as a science that is derived from the simultaneous application of electronics and pneumatics.

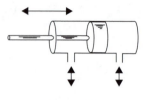

Figure 3.7

Formula:
(a) Forces

$$F_1/F_2 = S_1/S_2 \tag{3.11}$$

where F_1 = force applied to piston 1
F_2 = resultant force in piston 2
S_1 = area of piston 1
S_2 = area of piston 2

(b) Displacements

$$d_1/d_2 = F_1/F_2 \tag{3.12}$$

where d_1 = displacement of piston 1
d_2 = displacement of piston 2
S_1 = area of piston 1
S_2 = area of piston 2

Gas Equation

Ideal gases obey Eq. (3.13), which describes the state of rarefied gas in a container (see Fig. 3.8).

Formula:

$$PV = nRT \tag{3.13}$$

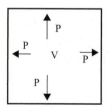

Figure 3.8

where P = pressure of the gas
        V = volume of the gas
        n = number of moles (mass/molecular mass)
        R = gas constant = 0.0821 atm·l/moles·deg

AIR MUSCLES

"Air muscles" are devices that can convert air pressure into mechanical force as shown in Fig. 3.9. In the figure, some applications for air muscles are shown: (a) arm, (b) gripper, and (c) actuator.

More information about air muscles can be found at www.imagesco.com/catalog/airmuscle/AirMuscle.html.

Chapter 10 describes some electronic circuits that can be used to control solenoid valves and electric pumps.

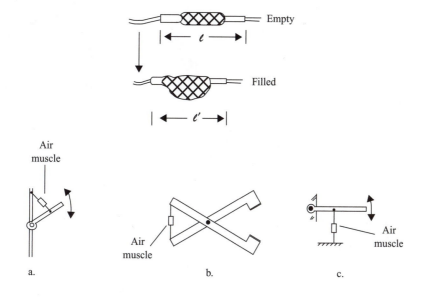

Figure 3.9

CHAPTER 4

Electric Motors

DC MOTORS

The simplest way to add motion to robots, mechatronic devices, and other automated equipment is through the use of DC motors. DC motors are cheap, small, and efficient, and they can be found in a wide range of sizes, shapes, and power ratings.

A conventional DC motor is formed by an arrangement of coils and magnets that create motion from electrical power. Figure 4.1 shows the typical construction of a small DC motor.

We can use these motors to move any mechanism directly or by adding gears or tracks to change the speed or increase the power (see Gearboxes, in Chapter 2).

When using a DC motor in a robotic or mechatronic project, the basic properties discussed below must be considered.

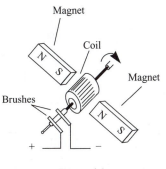

Figure 4.1

DIRECTION

When powered from a DC power supply (battery or other source), the direction the shaft rotates depends on the direction of the current flowing through it. By reversing the current, we can reverse the direction of any device driven by a DC motor.

SPEED

The speed of a motor (measured in rotations per minute, or rpm) depends on the voltage and the load.

We must consider two situations when using a DC motor. In the first case, the motor operates without any load or with a constant load. The speed will increase to a maximum that depends on the applied voltage.

In the second case, the motor operates with a variable load (that is, the motor powers some kind of mechanism whose forces depend on the moment or the task). In this case, the speed depends on the load: more power = less speed.

Figure 4.2 shows typical performance curves for these two cases: (a) no load and (b) under load.

VOLTAGE

Small DC motors can be found in voltages ranging from 1.5 to 48 V. The specified voltage indicates the nominal voltage or the applied voltage that makes the motor run in its normal condition (i.e., producing maximum power and consuming the nominal current). In practice, the nominal voltage is important in a project, as it indicates the maximum recommended operating voltage.

CURRENT

When a motor is powered at the nominal voltage, the current depends on the load. The current increases with the load. It is important not to allow the motor to run with excessive loads that can stall it. In the stalled condition, the motor represents a short circuit, and all of the applied power is converted into heat. The motor can burn out if left in this condition too long. Common DC motors have operating currents in the range of 50 mA to more than 2 A.

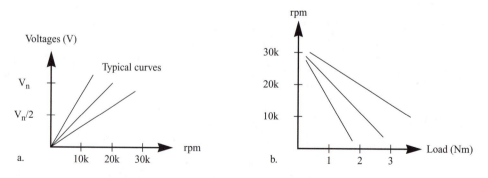

Figure 4.2

POWER

The power of a motor is given by the product of its voltage and current. In projects involving robotics and mechatronics, it is normal to rate the amount of force that a motor produce in terms of its *torque* (rotational power).

As shown in Fig. 4.3, torque is the force produced at the shaft, and it depends not only on the electrical and mechanical characteristics of the motor but also on the shaft diameter (see Chapter 2).

This specification is important, because the force that a mechanism powered by a DC motor can deliver depends not only on the motor but also on the mechanism coupled to it. Therefore, if gearboxes are added as shown in Fig. 4.4, the speed can be decreased, and the power can be increased to the same degree.

For example, if a gear with ten times the diameter of the shaft is coupled to a motor, the speed is reduced ten times, but the power is also increased by a factor of ten (see Chapter 2). Stated as an equation, if $D/d = 10$, then

$$\frac{V_1}{V_2} = 10 \text{ and } \frac{F_2}{F_1} = 10$$

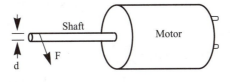

Figure 4.3

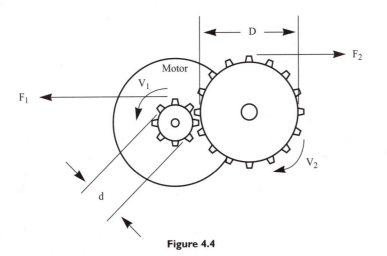

Figure 4.4

In robotics and mechatronics, the use of gearboxes of all sizes and reduction ratios is common for matching the characteristics of a particular motor to the desired task. Therefore, when using a motor, it is important to determine the torque; if we know the torque and the reduction rate of the driven mechanism, we can easily compute the final output power of the system.

SPEED

It is normal to rate the speed of a DC motor in terms of open (no load) conditions. Speeds can be in the range of 500 to 10,000 rpm, depending on the type, size, and other characteristics of any common DC motor. Remember that this speed will be reduced considerably when operating under load.

ADDITIONAL INFORMATION

Small, high-efficiency (HE) and low-efficiency (LE) DC motors and gearboxes can be obtained from many dealers. The most popular is Radio Shack, which can provide the designer with small motors and gear kits such as the following:

- 1.5 to 3 VDC, 8300 rpm, part number 273-223

- High speed, 12 V, 15,200 rpm at no load, part number 273-255

- Super speed, 9 to 18 V, 24,000 rpm at no load, part number 273-256

- Gear kit, two output shafts for different speeds and torque, ideal for use with motor 273-223, part number RSU 11903135

TYPICAL CHARACTERISTICS

The speed (rpm) of a small DC motors depends on the current and therefore the load. Typical values are shown in the curves of Fig. 4.5.

Observe that the speed increases when the amount of force generated by the motor is reduced, and so the current drawn by the unit is also reduced.

Operating the motor at an intermediate speed requires the use of electronic or mechanical controls. Pulse width modulation (PWM) controls, described later, can be used for this task (see p. 93).

PROTECTING THE DRIVE CIRCUIT

DC motors are inductive loads, and the action of the brushes can produce high-voltage spikes in the control circuit. These spikes can cause interference in other elements of the circuit or even cause damage in sensitive semiconductors such as CMOS devices and FETs. To avoid such problems, we can adopt the solutions shown in Fig. 4.6.

In Fig. 4.6a, we show how an electrolytic capacitor (between 10 and 1,000 µF) can be wired in parallel with the motor. If we intend to reverse the voltage to allow the

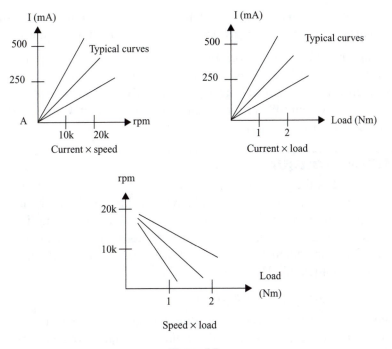

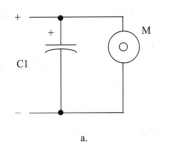

Figure 4.5

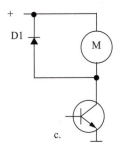

a.

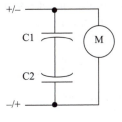

b.

c.

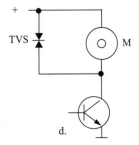

d.

Figure 4.6

motor to run in both directions, two electrolytics must be connected as shown in Fig. 4.6b. This configuration is equivalent to inserting a nonpolarized capacitor.

Keep in mind that this type of protection is not recommended if the motor is controlled by semiconductors such as bipolar transistors, power FETs, SCRs, and so on. These components do not like capacitive loads. If transistors or other semiconductors are used to control the motor, diodes and transient voltage suppressors (TVSs) must be used as shown in Figs. 4.6c and d.

MEASURING TORQUE

Torque is defined as the product of the force times the distance from the center of the shaft of a motor, as shown in Fig. 4.7. When using a DC motor (or any other motor), the main information the designer needs is the torque rating. If this information is not available, you must determine it through experimentation. Several simple techniques are suitable. Figure 4.8 shows some of the main techniques used to determine the torque of a DC motor.

a. Using a Prony brake * *dynamometer*

Since torque changes with speed, by loading the shaft as shown in Fig. 4.8a and using a dynamometer, it is possible to determine the torque. Notice that the force to load the motor is given by the operator. A device to measure the speed can be coupled to the shaft. (See the speed measurement techniques described in this chapter.)

b. Using a dynamo (generator) as a load

You can load the motor under test using the same dynamometer arrangement shown in Fig. 4.8a but substituting a simple generator (the type used with bicycles is suitable for small DC motors). The load is controlled by changing the current supplied to a lamp using a rheostat or other external resistance. Small DC motors can replace the dynamo, since they can operate as generators as well.

c. Using a known weight

This technique is suitable only for low-speed motors (gearboxes), since the measurement is based on the time delay in the displacement of the weight.

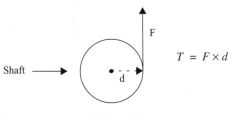

$$T = F \times d$$

Figure 4.7

*Named after its inventor, Gaspard de Prony.

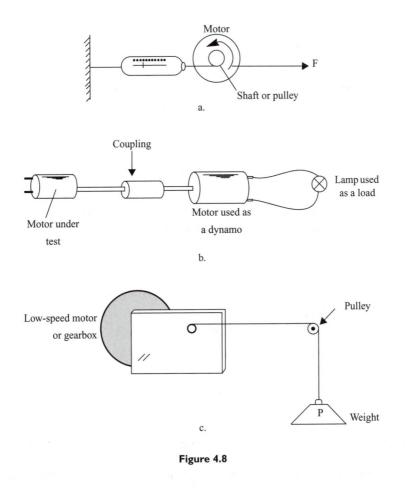

Motor

Shaft or pulley

a.

Coupling

Motor under
test

Motor used as
a dynamo

Lamp used
as a load

b.

Low-speed motor
or gearbox

Pulley

P Weight

c.

Figure 4.8

ESTIMATING TORQUE WITHOUT MECHANICAL MEASUREMENTS

Small DC motors have an efficiency[*] of between 60 and 80 percent when operating in their nominal speed ranges, i.e., 50 percent of the maximum speed. Using only electrical measurements with the circuit shown by Fig. 4.9, it is possible to estimate the torque of the motor using the following procedures:

1. Measure the current and voltage at the desired speed (use a load to limit the speed—refer to page 81 to see how the speed is determined).

2. Apply the following formulas.

Formulas:
Considering an efficiency of 80%,

$$T = 0.8\ P/\omega \qquad (4.1)$$

*Efficiency = measured torque as a percentage of theoretical torque.

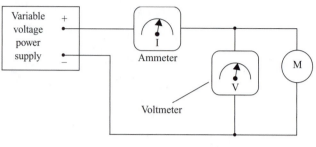

Figure 4.9

where P = power sink of the motor [as computed in Eq. 4.2)]
  ω = angular speed [as computed in Eq. (4.3)]
  T = torque

$$P = V \times I \qquad (4.2)$$

where P = power in watts
  V = voltage applied to the motor
  I = current drawn by the motor

$$\omega = 2 \times \pi \times f \qquad (4.3)$$

where ω = angular speed
  $\pi = 3.14$
  f = frequency (revolution per seconds)

Obs.: The same techniques can be used to estimate the torque of other rotational sources such as springs, hydraulic motors, and so forth.

SELECTING A HIGH-TORQUE DC MOTOR

The following are hints for finding a high-torque motor when rummaging through a store's surplus bin:

1. High-torque motors are heavy, due the presence of powerful magnets.

2. The high currents drawn by high-torque motors require large brushes.

3. The shafts tend to be large. Small-diameter shafts indicated less torque in common DC motors.

4. The shaft will resist being turned by your fingers, producing a jerky motion. This is not a conclusive test, but the probability is that a motor that "cogs" in a test with your fingers will have high torque.

5. The electrical connections are made with heavy-gauge wire. The high currents drawn by high-torque motors need heavier wires.

MEASURING SPEED

The speed of a DC motor is measured in revolutions per minute (rpm). As the designer should know, the speed depends on the load, applied voltage, and current.

In many applications, it is necessary to measure the speed of a motor when driving a load; in others, we measure the unloaded speed. There are several simple techniques that accomplish this task. Figure 4.10 shows the circuit and setup for this task.

(a) Using a motor as a dynamo

The voltage generated by a motor used as a dynamo depends on the speed of the motor under test. The motor used as a dynamo can be calibrated in terms of rpm using standard rotary sources.

(b) Using a stroboscopic LED

A stroboscopic light source is used to illuminate a stroboscopic disc. When motion is "frozen" by changing the frequency of the stroboscopic light source, it is possible to determine the rpm of the disc and therefore of the motor.

(c) Using an oscilloscope

Using a simple circuit that senses the presence of a shaft or a hole in the disc that is coupled to the motor under test, it is possible to precisely measure the rpm of a motor using an oscilloscope.

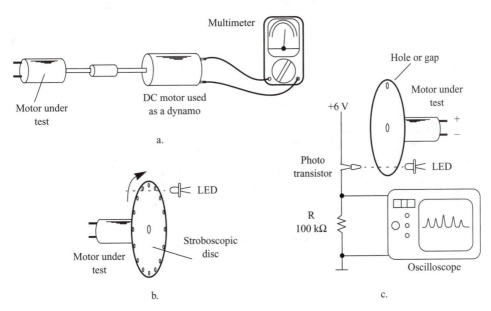

Figure 4.10 *(continues)*

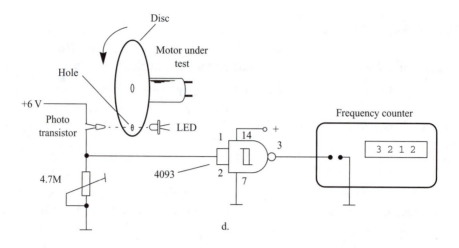

d.

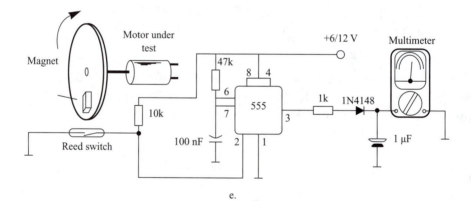

e.

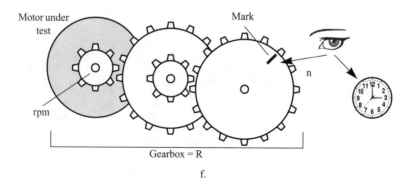

f.

Figure 4.10 *(continued)*

(d) Using a frequency meter

Figure 4.10d shows how the oscilloscope can be replaced by a frequency meter to determine the speed of a DC motor (or any other rotary device).

(e) Using a multimeter

The circuit shown in Fig. 4.10d converts the pulses produced by a magnet when passing near the reed switch into a proportional voltage. This voltage can be read by a common multimeter and associated with the speed of the motor. A calibration must be done first.

(f) Using a gearbox

By coupling the motor under test to a gearbox that provides a great deal of speed reduction, it is possible a visually count the revolutions of the final shaft. If you multiply this speed by the reduction rate of the gearbox, you can determine the speed of the motor under test.

Formula:

$$\text{rpm} = R \times n \tag{4.4}$$

where rpm = revolutions per minute of the motor under test
R = reduction rate of the gearbox
n = revolutions per minute of the gearbox shaft

H-BRIDGES

In this section, we will see how the half and full H-bridge can be used to control DC motors. H-bridges are very important for the robotics and mechatronics designers, as they simplify circuitry and allow the DC motor to be controlled directly from electrical signals, eliminating the need for relays or other mechanical parts as shown previously. H-bridges can also be used for controlling other polarity-sensitive loads such as solenoids and electromagnets.

THEORY

Changing the direction of a motor is a simple task if we use a relay with double-pole, double-throw (DPDT) contacts. But the use of relays presents some inconveniences: relays are not cheap; some are large, heavy devices; they have mechanical parts that are subject to failure; and it may be difficult to find one with the desired characteristics.

The use of transistors as switches, as discussed in Chapter 5, gives the designer additional resources for the control of DC loads. But, unfortunately, the control of a motor that must reverse its direction requires a DPDT switch, and a transistor operating as a switch is a single-pole, single-throw (SPST) device.

How do we solve this problem? The solution is given in this section. Two or four transistors can be linked in some special configurations to allow them to control a DC motor as a DPDT switch.

Two special configurations are used for this task: the half bridge and the H-bridge. Lets see how they work and how can we use them in projects involving robotics and mechatronics.

HALF BRIDGE

Two batteries and two SPST switches are used in the circuit illustrated by Fig. 4.11. In normal operation, two conditions are possible:

1. If SW1 is closed, the current is provided by B1, and the motor runs forward.

2. If SW2 is closed, the current flows across the motor in the opposite direction, now powered by B2, and it runs backward.

It should be clear that closing SW1 and SW2 at the same time is a forbidden condition, as it short-circuits the batteries.

If we replace the switches with transistors (bipolar or FET), we can control the motor by biasing the transistors from external voltage sources as shown in Fig. 4.12.

The polarity of the control voltage depends on the type of the transistor. NPN and P-channel power FETs are made conductive when positive voltages are applied to their bases or gates. On the other hand, PNP transistor are made conductive when negative voltages are applied (or they are grounded).

The main disadvantage of this configuration (using only two transistors) is that we need a dual or symmetric power supply. This configuration increases the degree of complexity of the drive circuits. The solution is to use a configuration of four transistors as discussed next.

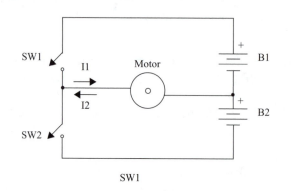

SW1

Figure 4.11

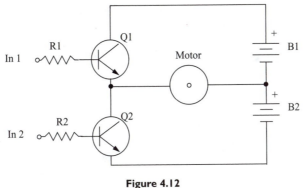

Figure 4.12

H-BRIDGE OR FULL BRIDGE

Let's now start from the basic circuit in which we use four SPST switches as shown in Fig. 4.13. In operation, we can encounter two situations:

1. Closing SW1 and SW4 causes the current to flow in one direction, and the motor runs forward.

2. Closing SW2 and SW3 causes the current to flow in the opposite direction, and the motor runs backward.

It is obvious that SW1 and SW3 can't be closed at the same time, since this will short-circuit the batteries. For the same reason, SW2 and SW4 can't be closed simultaneously.

Now replace the switches with bipolar or FET transistors as shown in the Fig. 4.14. Depending on the bias, we can make the transistors conduct or not conduct current, thereby acting as switches.

In this case as well, we must avoid the condition in which Q1 and Q3, or Q2 and Q4, are simultaneously conducting, as either will short out the battery.

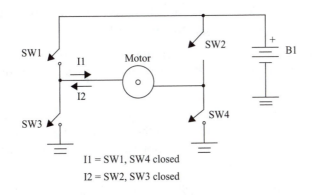

Figure 4.13

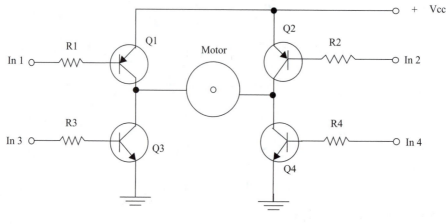

Figure 4.14

To correctly control this bridge we have two possibilities:

1. We can directly stimulate the transistors from sensors, switches, or other circuits in a configuration that is "intelligent" enough to avoid the forbidden conditions.

2. We can add "intelligence" or logic to the bridge to avoid the forbidden states.

In the second case, the "intelligent" configuration depends on the type of transistors used. If using a PNP and an NPN transistor, for example, it is sufficient to wire the bases of each transistor pair together, thereby biasing them according to the direction the motor must run.

If only one type of transistor (NPN bipolar transistors or power MOSFET P-channel) is used, inverters must be added to the circuit to provide the logic necessary to avoid forbidden conditions.

In all of these cases, we see that the signals must be applied to the transistor in a manner such that neither Q1 and Q3 nor Q2 and Q4 will conduct at the same time.

These two possibilities bring us to several types of circuits for DC motor control, using a half bridge or full bridge, depending on the signal sources or control voltage sources.

In Fig. 4.15a, we have only one input, and the motor runs in one direction when the input is high (positive voltage) and runs in the opposite direction when the input is low (ground). Notice that this configuration can be made compatible with such logic devices as TTL and CMOS.

The other option, shown in Fig. 4.15b, uses two signal sources. In this case, we have four possibilities of operation as shown in the following table:

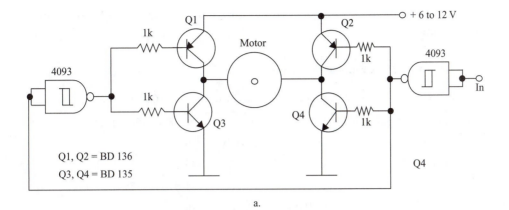

a.

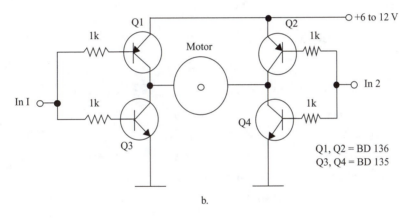

b.

Figure 4.15

In 1	In 2	Motor
High	Low	Runs forward
Low	High	Runs backward
Low	Low	Does not run
High	High	Forbidden condition

This combination of inputs can be varied to suit the configuration, regardless of whether we use only NPN transistors, only PNP, or NPN and PNP in the same bridge.

TRANSISTORS USED IN H-BRIDGES

The transistors that should be used in a half or full bridge depends on the current they must control. This current is determined by the motor and can vary from a few milliamperes to more than 2 A.

In the following tables, we will list some transistors that are suitable for applications in bridges.

Table 4.1   NPN Bipolar Transistors

Type	Voltage (Vce)	Current (Ic)	Gain (hFE)
BC547/548/549	45/30/30 V	100 mA	110–800/110–800/200–800
2N3903/3904	40 V	200 mA	20–200
BD135/137/139	40/60/80 V	1.5 A	40–250
BD433/435/437	22/32/45	4 A	85–475/85–475/85–375
TIP31/A/B/C	40/60/80/100 V	3 A	10–50
TIP41/A/B/C	40/60/80/100 V	6 A	15–75

Table 4.2   PNP Bipolar Transistors

Type	Voltage (Vce)	Current (Ic)	Gain (hFE)
BC557/558–559	45/30/30 V	100 mA	75–800/75–800/125–800
2N3905/3906	40 V	200 mA	20–200
BD136/138/140	40/50/80 V	1.5 A	40–250
BD434/436/438	22/32/45 V	4 A	85–475/85–475/85–375
TIP32/A/B/C	40/60/80/100 V	3 A	10–50
TIP42/A/B/C	40/60/80/100 V	6 A	15–75

Table 4.3   NPN Darlington Transistors

Type	Voltage (Vce)	Current (Ic)	Gain (hFE)
BD331/333/335	60/80/100 V	6 A	750
TIP110/111/112	60/80/100	1.25 A	500
TIP120/121/122	60/80/100	5 A	1000
TIP140/141/142	60/80/100	10 A	1000

Table 4.4   PNP Darlington Transistors

Type	Voltage (Vce)	Current (Ic)	Gain (hFE)
BD332/333/335	60/80/100 V	6 A	750
TIP115/116/117	60/80/100 V	1.25 A	500
TIP125/126/127	60/80/100 V	5 A	1,000
TIP145/146/147	60/80/100 V	10 A	1,000

Table 4.5 Power MOSFETs (N–channel)

Type	Voltage (Vds)	Current (Id)	Rds (on)
IRF222	200 V	4.0 A	1.2 Ω
IRF230	200 V	9.0 A	0.4 Ω
IRF250	200 V	30 A	0.085 Ω
IRF223	150 V	4.0 A	1.2 Ω
IRF231	150 V	9.0 A	0.4 Ω
IRF122	100 V	7.0 A	0.4 Ω
IRF142	100 V	24 A	0.11 Ω
IRF123	60 V	7.0 A	0.4 Ω
IRF143	60 V	24 A	0.11 Ω
IRF151	50 V	40 A	0.055 Ω

SPECIAL RECOMMENDATIONS

Decoupling Capacitors

When switching an inductive load (e.g., a DC motor) on and off, high-voltage and high-current spikes are generated that can propagate across the circuit and cause electromagnetic interference. To avoid such problems, much care must be taken by the robotics and mechatronics designer.

Decoupling capacitors must be included wherever current or voltage spikes could be present and compromise the stability of the circuit. Some circuit locations are particularly well suited for these capacitors.

- *In parallel with the power supply.* Large electrolytic capacitors should be added in parallel with the power supply as shown by Fig. 4.16. These capaci-

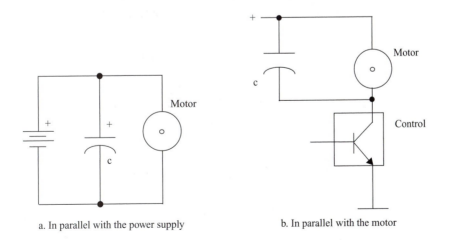

a. In parallel with the power supply   b. In parallel with the motor

Figure 4.16

tors and add stored energy to the current supplied by the source when the motor is switched on and momentarily draws high current.

Values between 1,000 and 10,000 µF are suitable for common applications. If the motors draw very high currents when switched on, and a drop in the supply voltage can't be avoided, it is best to use a separate battery for each motor.

• *In parallel with the motor.*   Capacitors in the range between 0.1 and 1 µF (polyester types) are recommended for absorbing the spikes generated by the motor's commutation system. These capacitors will protect the switching devices against high-voltage spikes that are produced when the motor is running.

Protecting the Transistors

Transistors are very sensitive to voltage spikes that travel across them; e.g., spikes generated by inductive loads. Many types of transistors used for switching include internal protective diodes. If the transistors that you use do not have them, it is recommended that you add protection as shown in Fig. 4.17.

The ideal devices for this purpose are Schottky diodes, since these devices switch quickly enough to deal with the short pulses generated by DC motors. But, for common applications, your circuits can be protected with common diodes such as the 1N4002 and others in the same series.

Current Sensing

In some applications, it is important to add a circuit to monitor the current drain of a motor. This circuit can be used to control speed or manage power use to extend battery life. Figure 4.18 shows how this circuit can be added.

A 0.1 Ω resistor causes a voltage drop of 100 mV for each ampere drawn by a motor. Adding an operational amplifier with a gain of 10, for instance, you obtain 1 V/A directly driving some control circuit or monitor circuit. The value of the resistor must be kept as low as possible to avoid "robbing" energy from the motor.

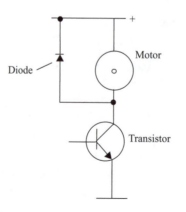

Figure 4.17

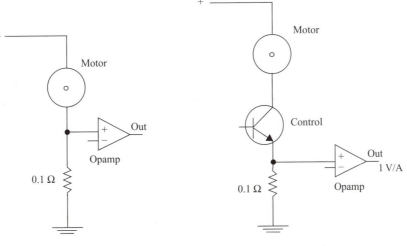

Figure 4.18

Fuses

Many blocks shown in this section have "forbidden states" that can cause a power supply short out. The high current that would pass across the transistors in such conditions could burn them out.

To protect the circuit, fuses are recommended. The fuses are placed in series with the power supply, and their values depend on the motors and the transistors. It is a good rule of thumb to use fuse values 2.5 to 3 times greater than the normal current drawn by the motors.

LINEAR AND PWM POWER CONTROLS

Controlling the speed of a motor is very important in many robotics and mecha-tronics projects. Below, we provide some basic information about techniques for the control the speed of a DC motor. The same principles can be applied to control other loads such as the force of an electromagnet or solenoid, the temperature of a heater, or the brightness of an incandescent lamp.

BASIC THEORY

The simplest way to control the power applied to a load is by installing a rheostat in series as shown in Fig. 4.19. The rheostat and the load form a variable voltage divider. By changing the resistance of the rheostat, the voltage applied to the motor is also changed and, consequently, so is its speed.

This type of control, although very simple, presents some inconveniences.

1. The current drawn by the motor also passes across the rheostat. Depending on the size and power of the motor, this means that a large amount of heat may be

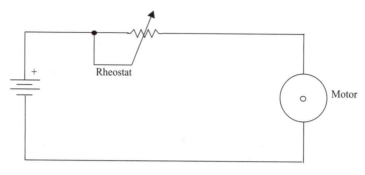

Figure 4.19

generated in the rheostat. In this case, special types of heat dissipation devices must be used, which increases the size and the cost of the component.

2. When controlling a DC motor, since it also represents a variable load (the current across a motor depends on the load and speed), the circuit is unstable; it is difficult to make the motor start smoothly, maintaining constant torque. The tendency is for a hard start, making the robot or other application jerk forward.

These two annoyances can be eliminated through the use of the special configurations described in this section. As in other parts of the book, the blocks are basic and must be altered to match their characteristics to your applications and motors. All the information needed to perform this task is given in the following pages.

TWO TYPES OF CONTROLS

Two circuit configurations can be used to control the power applied to a motor or other load, using modern components to avoid the previously described problems. (A third kind of control will be described later; it uses special components that are part of the device's core.)

LINEAR CONTROLS

The linear control uses a transistor (bipolar or FET) as a variable resistor in a typical configuration as shown in Fig. 4.20.

The base (or gate) current changes the resistance between collector and emitter and therefore the current flowing across the circuit.

The main advantage of this circuit is that the base current is low when compared with the current flowing across the transistor, meaning that this component doesn't need to dissipate much power. The transistor is the element that will dissipated the power. This allows the use of common carbon low-dissipation potentiometers in the control of high current loads.

The disadvantage is the one discussed before: we have power losses, because the transistor converts a considerable amount of power into heat as it controls the current across the load.

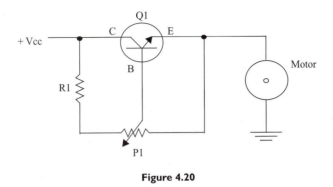

Figure 4.20

The power converted into heat is computed by the voltage drop across the transistor times the current controlled.

For instance, if a transistor is adjusted to apply 6 V to a 0.5-A motor from a 12-V power supply as shown in the figure, the voltage drop of 6 V across the transistor times the current (0.5 A) will result in the production of 3 W of heat.

PULSE WIDTH MODULATION

Pulse width modulation (PWM) is a much more efficient power control technology. As a result, it is used not only to control DC loads but also in many other applications (e.g., power supplies).

The basic idea starts with the use of square voltage pulses to power a load. The amount of power applied to the load depends on the duration of each pulse or the duty cycle of the signal.

If the duration of the pulse is the same as the interval between pulses, representing a duty cycle of 50% (see Fig. 4.21a), the average power applied to the load is 50%. Extending the duration of the pulse, the average power applied to the load increases in the same proportion, as shown by Fig. 4.21b.

By controlling the "size" ("width") of the pulses, we can control the power applied to a load. The process used to control the width of the pulse is called *modulation,* so this kind of circuit is called a *pulse width modulation power control* or *PWM power control.*

How does a practical PWM control circuit work? Let's start with the configuration shown in Fig. 4.22.

A power transistor (MOSFET or bipolar) is wired to the output of a variable-duty-cycle oscillator. When the oscillator is running, the transistor turns on and off at the oscillator frequency, applying a square voltage to the load. The average voltage in the load, as we have seen, depends on the duration of the pulses. The great advantage of this type of circuit is that the power dissipated by the transistor is nearly zero.

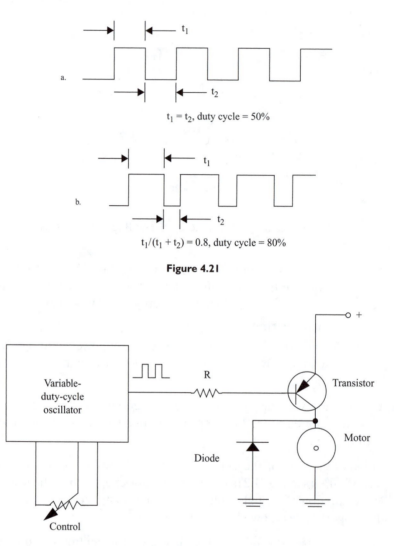

Figure 4.21

Figure 4.22

When the transistor is on, its resistance can be considered zero, and no power is generated across it (i.e., the power is the product of the voltage drop times the current and, since the voltage drop is practically zero, the result is zero power). On the other hand, when the transistor is off, no current flows through it, and again the power dissipated is zero.

When working with "real-world" components, things are quite different. The transistor can't pass from the on state to the off state, or from the off state to the on state, in a short enough time to avoid dissipation problems. The transistor needs a finite time for this state change.

During the time it takes the transistor to switch from one state to the other, the resistance presented by the device changes; at the same time, current and voltage drops appear in the circuit. These voltages and currents are large enough to cause some heat to be generated.

Although the heat generated by the switching process can be important in this kind of circuit, it is very low when compared to the heat generated by linear controls, as the voltage and current drops are present only part of the time.

Another important advantage of PWM circuits for controlling DC motors is that they maintain more constant torque throughout the entire speed range. As we have seen, when using a linear control, the DC motor tends to jump forward when it draws the power needed to overcome inertia. Using a PWM control, all of the pulses have the total circuit voltage; only the duration changes.

This means that, even at very low speeds, the motor receives enough voltage to overcome inertia and begin to rotate. From this level, it is possible to adjust the speed smoothly throughout the speed range, which is not possible with a linear control.

TWO FORMS OF PWM CONTROLS

Two forms of PWM controls can be found in practical applications involving robotics and mechatronics.

Locked Anti-phase Control

The simple locked anti-phase control consists of a single, variable-duty-cycle signal oscillator in which is encoded both direction and amplitude (speed) information, as shown in Fig. 4.23.

A 50 percent duty-cycle signal represents zero drive, since the net voltage value (integrated over one period) delivered to the load (motor) is zero. The great disadvantage to this kind of control is that, when the applied power is zero, the power supplies are producing power 50 percent of the time. This power is converted to heat. For that reason, this type of control is not recommended for use with high-power motors.

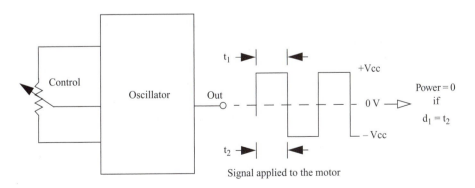

Figure 4.23

Sign/Magnitude Control

This control consists of separate direction (sign) and amplitude (magnitude) signals. The magnitude signal is duty-cycle modulated, and the absence of a pulse signal (a logic high level) represents zero drive. Figure 4.24 shows what happens with this control.

Current delivered to the load is proportional to the pulse width. Figure 4.25 shows some simple circuits for linear and PWM control of DC motors: (a) simple linear circuit using a bipolar transistor, (b) simple linear control using a Darlington transistor, (c) linear control using a voltage regulator IC (LM350T), (d) PWM using a 555 IC, (e) PWM anti-phase using a 555 IC, and (f) PWM using a 4093 IC.

STEPPER MOTORS

Stepper motors may be used for locomotion, movement, positioning, and many other functions where we need precise control of the position of a shaft, lever, or other moving part of a mechatronic device. The purpose of this section is to provide basic information about the use of stepper motors in mechatronics and robotics projects, and to show circuit blocks that can be used for practical applications.

BASIC THEORY

The basic operating principle of a stepper motor is not much different from that of a DC or AC motor. Steppers are formed by coils and magnets and incorporate a shaft that moves when power is applied. The difference is in the way the shaft is moved: they move the rotor by applying power to different coils in a predetermined sequence (i.e., they are *stepped*). Steppers are designed for fine control requirements. They will spin on command, but they will also turn at a particular number of steps per second up to the maximum speed. Another function that makes them more versatile than common motors is that stepper motors can hold their position and resist rotation.

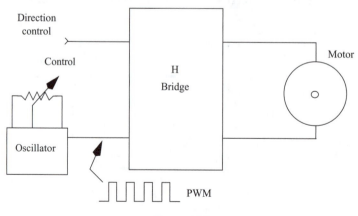

Figure 4.24

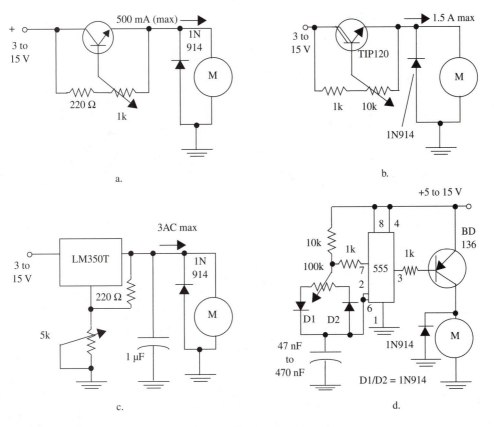

Figure 4.25 *(continues)*

Figure 4.26 shows the symbol adopted to represent a stepper motor and aspects of the most common types.

Robotic and mechatronic designers don't need to buy expensive stepper motors for their projects, since many can be salvaged in good condition from computer disk drives, printers, and many other discarded or surplus devices.

HOW IT WORKS

Stepper motors convert digital information into proportional mechanical movement. They are different from DC motors that are controlled by changing the current across them. Stepper motors are digital.

Stepper motors are available in three basic types: permanent magnet, variable reluctance, and hybrid.

The way the windings are organized inside a motor determines how it works. The most common type is the four-phase stepper motor, but there are also two-phase and six-phase types.

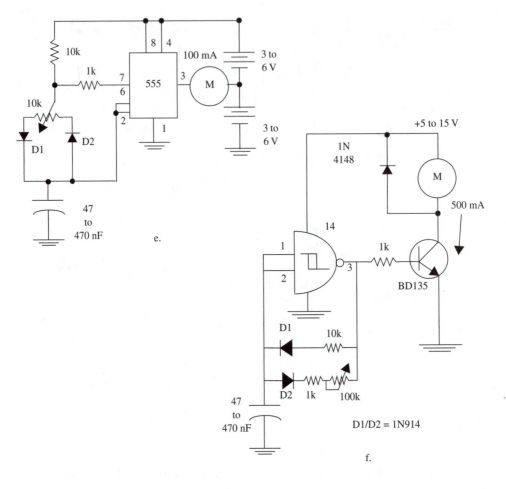

Figure 4.25   *(continued)*

Figure 4.27 shows the most popular of the stepper motor types: the four-phase stepper motor. Inside this motor, we find four windings. Since each pair of windings has a common connection, this kind of motor can be easily identified by the six wires used for connections.

In normal operation, the common wires are connected to the positive line of the power supply, and the other wires are attached to ground for a short period of time; i.e., for as long you want to energize the corresponding winding.

Each time the motor is energized, the motor shaft advances a fraction of a revolution. For the shaft to turn properly, the windings must be energized by a sequence of pulses or waves. For instance, if you energize windings A, B, C, and D in that sequence, the shaft turns clockwise. On the other hand, if you reverse the sequence, the motor turns counterclockwise.

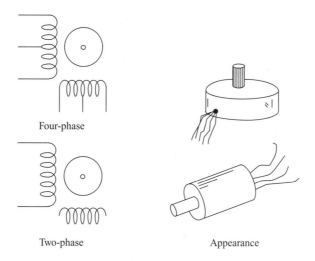

Four-phase

Two-phase

Appearance

Figure 4.26

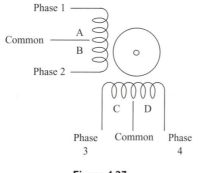

Phase 1

Common

A

B

Phase 2

C   D

Phase
3

Common

Phase
4

Figure 4.27

Figure 4.28 shows the sequence that is normally used to energize four-phase stepper motors. Another way to energize a stepper motor is by applying an on-on/off-off sequence. This sequence is shown in Fig. 4.29. An advantage of this sequence is that it increases the drive power of the motor and provides greater shaft rotation precision.

Another common type of stepper motor is the two-phase design, shown in Fig. 4.30. This stepper motor is formed by two coils, as shown in the figure, and it can be easily identified by its four wires. This type of motor is energized using a different sequence in which the direction of the current across each winding is considered, as shown in Fig. 4.31. In some types, the –V can be replaced by ground.

There are also stepper motors with additional phases, such as the six-phase stepper motor, but they not very common. Motors with a greater number of phases are more precise but also more expensive.

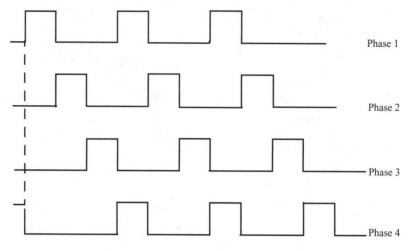

Figure 4.28

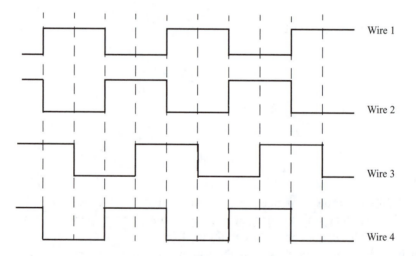

Figure 4.29

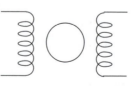

Figure 4.30

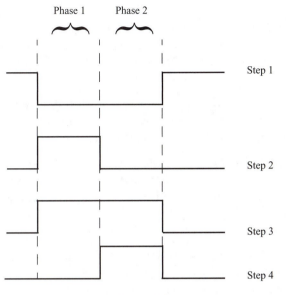

Figure 4.31

For our purposes (i.e., various applications in robotic and mechatronic projects), the four-phase stepper motor is recommended, and most of the blocks in this chapter are designed to employ this type of motor.

HOW TO USE STEPPER MOTORS

As we have seen, the windings of a stepper motor must be energized properly to achieve correct operation. This means that, when using a stepper motor, you'll need to know not only the electric specifications for this kind of device but also the mechanical specifications.

The following are the most important.

VOLTAGE AND CURRENT

In most cases, stepper motors have voltage ratings of 5, 6, or 12 V. Unlike conventional DC motors, is not recommended that you overdrive the windings of a stepper motor. Overvoltages of more than 30 percent of the rating can burn the windings.

The current ratings depend on the application (size and torque). Common types can draw currents in the range from 50 mA to more than 1 A. The higher the current and voltage, the higher the torque.

When designing a power supply for an application using a stepper motor, it is important to consider that the current ratings are given per winding. So the power supply must be able to supply at least double the current per winding, or eight times the current per winding in a four-phase type.

SEQUENCE

Although most stepper motors use one of the two sequences described previously, some units use different sequences. When using such units, you must determine the sequence of pulses that produces correct operation.

STEP ANGLE

When a pulse in the sequence is applied to the motor, it advances one step. This means that the shaft moves a predetermined number of degrees. Depending on the type of motor, this can vary in the range between 0.8° and 90°.

In a 90° stepper motor, four pulses drive the shaft to complete one turn, as shown in Fig. 4.32. But it is more common to find stepper motors with step angles of 1.8°. This means that you must apply 200 pulses to the control circuit to drive the motor one complete turn.

PULSE RATE

The pulse rate determines the speed of the motor. If you are using a 1.8° step angle motor, and you apply 200 pulses per second, the motor will run at 1 rotation per second or 60 rotations per minute (60 rpm). Given the step angle, it is easy to calculate the rpm.

Stepper motors are not intended for high-speed applications. The top recommended speeds are in the range of 2 or 3 turns per second, or 120 to 180 rpm. It is important to remember that, in this kind of motor, the torque drops as the speed increases.

TORQUE

The torque produced by a stepper motor is not very high. A typical stepper motor can provide only a few grams per centimeter when running. This means that, in applications where high torque is needed, gearboxes must be added.

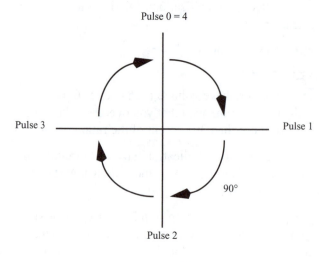

Figure 4.32

Since torque falls with speed, it is important to use these motors in low-speed modes. Refer to page 78 for information about torque measurement.

BRAKING EFFECT

If you maintain the current flow in a winding after a pulse has been applied, the stepper motor can't continue to turn. The shaft becomes locked as if you applied a brake. A circuit that maintains the current in a winding to hold a particular position is, in effect, an electronic brake for the stepper motor.

Figure 4.33 shows the following circuits used to control stepper motors:

Figure 4.33a. Using bipolar transistors, current up to 500 mA for each winding

Figure 4.33b. Using Darlington transistors, current according to the transistor

Figure 4.33c. Using power FETs, current according to the transistor

Figure 4.33d. Using the ULN2003, four outputs from the PC, which generates the sequence of pulses, each output controls one winding

Figure 4.33e. Using the ULN2003, two outputs from the PC, which generates the sequence in binary form (00, 01, 10, and 11)

Figure 4.33f. Switch controller using the ULN2803

Figure 4.33g. Serial controller using the 12C509/ULN2803

ICS RECOMMENDED FOR STEPPER MOTORS DRIVERS

The following table provides information about other transistor arrays and power stages that are recommended for stepper motor drive circuits.

Type	Current	No. stages	Manufacturer	Observations
ULN2003	500 mA	7	*	16 pins
ULN2803	500 mA	7	*	Equivalent to the UCN2003 but in 18-pin package
MC1413	500 mA	7	Motorola	TTL compatible
MC1414	500 mA	7	Motorola	CMOS compatible
SN75465	500 mA	7	Texas Inst.	1050 Ω series resistor at each input, TTL compatible
SN75466	500 mA	7	Texas Inst.	2700 Ω series resistor at each input, CMOS compatible
SN75468	500 mA	7	Texas Inst.	Indicated for CMOS and TTL with 5 V supply
SN75469	500 mA	7	Texas Inst.	10.5 kΩ series resistor at each input, CMOS with 6 to 15 V supplies

* Available from Jameco Electronics, www.jameco.com.

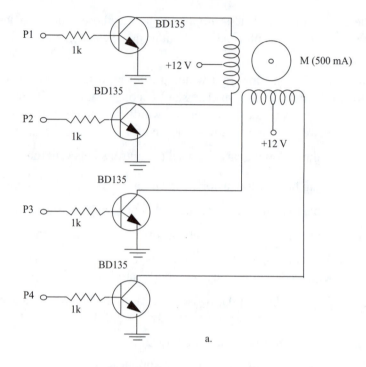

a.

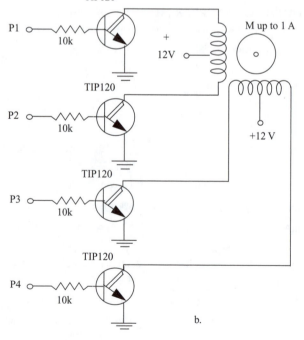

b.

Figure 4.33 *(continues)*

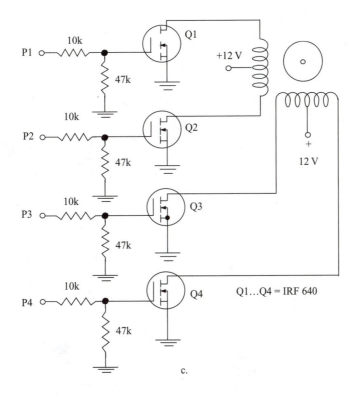

c.

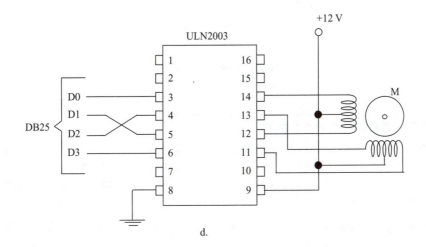

d.

Figure 4.33 *(continued)*

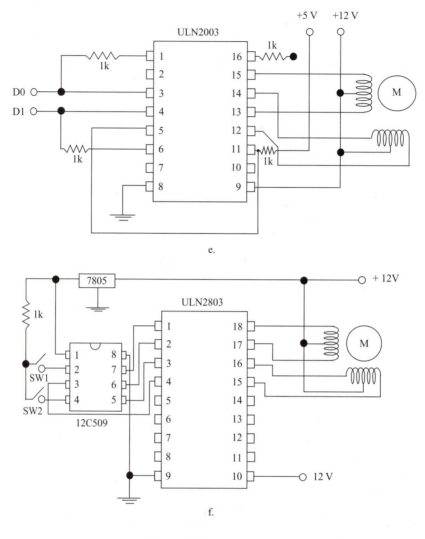

Figure 4.33   *(continued)*

PIEZOELECTRIC MOTORS

Piezoelectric motors are used in applications in which we need fast response times, high precision, hard braking, backlash elimination, and high power-to-weight ratios. Unfortunately, they are neither easy to use nor easy to find for our applications, since they are expensive.

Even though advancements in manufacturing technologies have resulted in lower costs, these motors still are not very popular. It is likely that, within a few years, piezoelectric motors will become as popular as the more familiar types.

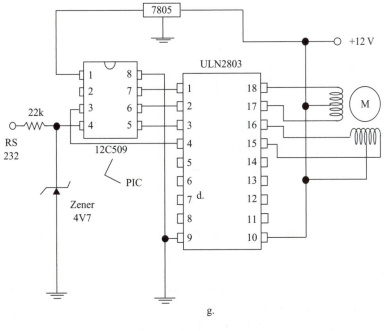

Figure 4.33   *(continued)*

Some applications for piezoelectric motors are as follows:

1. Micropositioning stages

2. Manufacturing process control

3. Fiber optic positioning

4. Camera autofocus mechanisms

5. Computer disk drives

6. Robotic positioning

HOW THEY WORK

Piezoelectric materials generate an electric field when stressed along a primary axis. Conversely, the material changes its shape if a voltage is applied along this axis.

A common piezoelectric material is lead zirconate titanate. This type of ceramic material, as shown in Fig. 4.34, can be polarized by high-voltage electric fields. Once the domains are properly aligned, the piezoelectric material can do useful work.

Piezoelectric motors use a polar ceramic shape to create motion through the use of periodic (sinusoidal) electric fields.

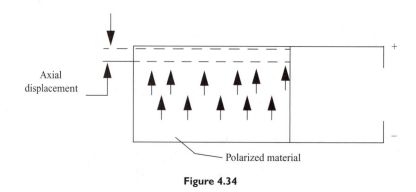

Axial
displacement

Polarized material

Figure 4.34

The typical piezoelectric characteristic of a ceramic material is 500 ppm/V (parts per million per volt), meaning a displacement of only 0.1 microns if 200 V is applied.

Piezoelectric motors can be classified as standing-wave (a single vibration driving source) or propagating-wave (two standing waves with a phase shift of 90°) types. Figure 4.35 shows the main types of piezoelectric motors. The basic construction of a piezoelectric motor is shown in Fig. 4.36.

The force along the wear strip is about 2 to 3 N. The force can be calculated by the following equations.

Formulas:

$$V = V_o \times \sin (\omega \cdot t) \tag{4.5}$$

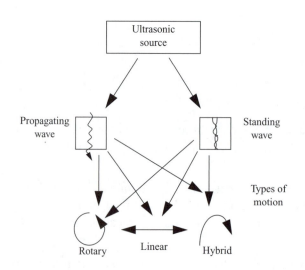

Figure 4.35

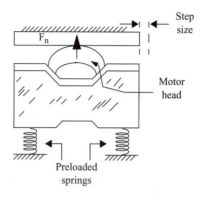

Figure 4.36

and

$$F = \mu \times F_n \tag{4.6}$$

where V_o = the original peak voltage (V)

f = the frequency (Hz)

t = the time (s)

F = the kinetic force (N)

μ = the coefficient of sliding friction

$\omega = 2 \times \pi \times f$

F_n = the normal force (N)

CHAPTER 5

Relays

THE RELAY

Relays are electromechanical switches. A relay is basically formed by a coil and one or more pairs of contacts as shown in Fig. 5.1.

When a voltage is applied to the coil, the current flow creates a magnetic field that attracts the contacts and closes the switch. If the current across the coil is cut, the magnetic field disappears, and the contacts open. An external circuit controlled by the contacts can be triggered on and off by the current passing across the coil.

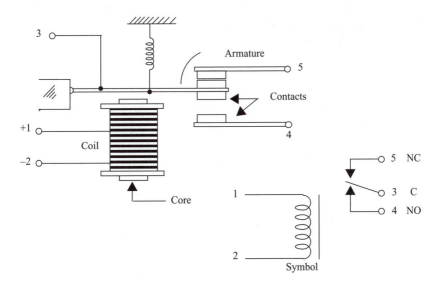

Figure 5.1

Two important properties can be noted in this arrangement:

1. The controlled circuit is completely isolated from the control circuit.

2. We can apply low voltage and low current to the relay's coil to control high-voltage and/or high-current circuits.

Relays come in different sizes and with various electrical characteristics as needed for a particular application. The main characteristics we must consider when using relays are as follows:

1. *The nominal voltage or coil voltage (V_{coil}).*   This is the voltage that is applied to the coil to make the contacts close. In practice, the relay can close its contacts with voltages lower than indicated, and keep them closed even when the voltage falls below the nominal value, as a result of the characteristic of *hysteresis*. Types with voltage ratings between 3 and 48 V are common in robotics and mechatronics applications. Figure 5.2 illustrates.

2. *The coil current (I_{coil}).*   This is the current that flows across the coil when the nominal voltage is applied. Currents between 20 and 100 mA are common in relays.

3. *The coil resistance (R_{coil}).*   This resistance can be found easily by dividing the nominal voltage by the coil current. Values between 50 and 500 Ω are common.

4. *Contact current ($I_{contact}$).*   This is the maximum current that can be controlled by a relay. Typical values are in the range from 1 to 10 A.

5. *Type of contacts.*   The relays are not merely simple switches controlled by currents. One coil can control one or more contacts in different configurations,

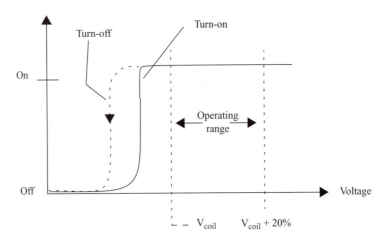

Figure 5.2

as shown in Fig. 5.3. This figure also shows the symbol and the appearance of common relays.

In Fig. 5.3a, we have a single-pole, single-throw (SPST) relay, and in Fig. 5.3b a DPDT relay. Other types with multiple contacts can be obtained. Therefore, relays can act as SPST, DPDT, or multiple-contact/throw switches.

With the following formulas, you can compute any third characteristic of a relay when the other two are known:

Formulas:

1. Resistance of the coil when voltage and current are known,

$$R = V/I \qquad\qquad (5.1)$$

2. Current in the coil when resistance and voltage are known,

$$I = V/R \qquad\qquad (5.2)$$

3. Voltage in the coil when current and resistance are known,

$$V = R \times I \qquad\qquad (5.3)$$

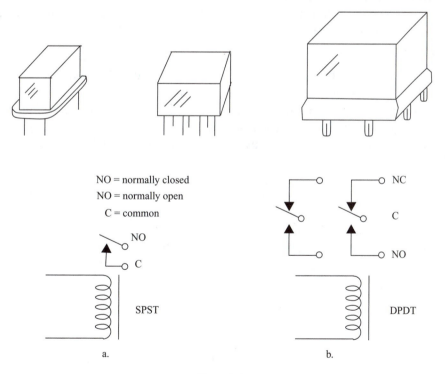

Figure 5.3

In some cases, it is important to know the power dissipated by a relay in an application. The following formula can be used for this task:

Formula:

$$P = V \times I = R \times I^2 = V^2/R \qquad (5.4)$$

where P = power dissipated by the relay coil (W)
V = applied voltage (V)
I = current through the coil (A)
R = resistance of the coil (Ω)

SPECIAL TYPES OF RELAYS

Latch Relays

Latches or *bistable relays* are turned on by a current flowing in one direction, and they maintain the "on" state even when the current is cut. To turn off the relay (or to bring it back to the previous state), it is necessary to apply another current to the coil—but in the opposite direction. Some latch relays use two coils—one to turn on and the other to turn off.

Solid State Relays

These are special types in which the contacts are replaced by some kind of solid state switch such as a transistor, SCR, TRIAC, or other device. More information about these devices can be found in Chapter 10.

Reed Relays

These are relays formed by a reed switch and a coil in the configuration shown in Fig. 5.4. The operating principle is the same as with common relays: when a current flows through the coil, the resulting magnetic field acts on the contacts of the reed switch and closes them. Reed relays are small and very sensitive but can't handle large currents. Normally, they are used to control currents up to several hundred milliamperes.

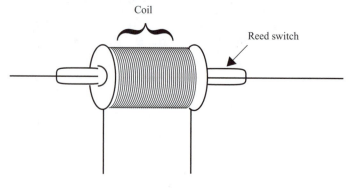

Figure 5.4

HOW RELAYS ARE USED

When using a relay, we have to consider both the device that is to be controlled (wired to the contacts) and the circuit that will drive the relay. Both are discussed in the following sections.

USING THE CONTACTS

Figure 5.5 shows how relay contacts are configured. In this figure, we show several applications as described below:

Figure 5.5a. Here we show the simplest application of the relay. It is used as an SPST switch to control an external load. The load is on when current flows across the coil.

Figure 5.5b. Here we show how the normally closed (NC) contacts can be used to turn a load off when the relay is energized. This configuration is preferred when the load's on time exceeds the off time.

Figure 5.5c. Figure 5.5c shows how a DPDT relay can be used to control the direction of the current flowing across a load. When the relay is off, the direction of the current is I_1. When the relay is on, the direction is I_2.

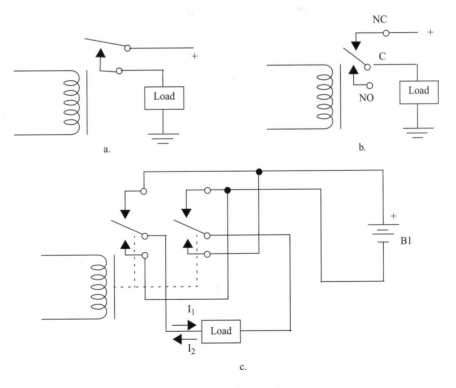

Figure 5.5 *(continues)*

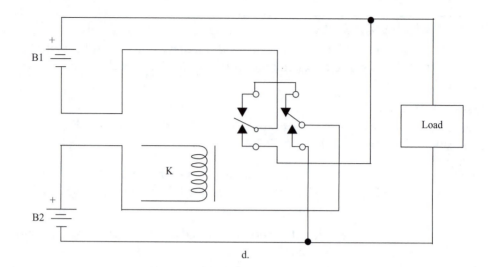

d.

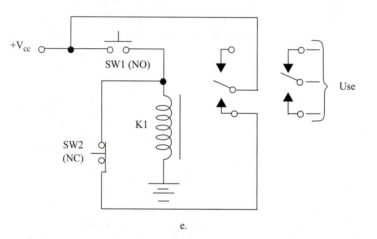

e.

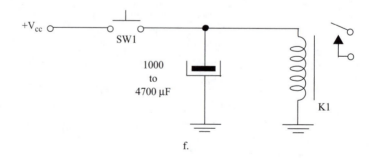

f.

Figure 5.5   *(continued)*

Figure 5.5d.   In Fig. 5.5d, a relay is used to switch power supplies or loads, changing from series to parallel and vice versa.

Figure 5.5e.   Another interesting application is shown in Fig. 5.5e. When SW1 is pressed, one pair of contacts is used to latch the relay. To turn off the relay, we momentarily press SW2. Notice that SW1 is a momentary, normally open switch, and SW2 is a momentary, normally closed switch.

Figure 5.5f.   As shown in Fig. 5.5f, by momentarily pressing SW1, the relay closes the contacts and remains in this state for an interval that is determined by the value of the capacitor and the resistance of its coil. A sensor can be used as SW1.

DRIVE CIRCUITS USING SENSORS AND SWITCHES

Figure 5.6 shows a variety of circuits that can be used to drive relays from low-power on/off sensors or low-power switches such as reed switches, mercury switches, blade sensors, pendulum sensors, and others. The current flowing through the sensor/switch is the current used by the coil to close the relay contacts.

Let's explain the operation and use of each configuration.

Figure 5.6a.   This is the simplest configuration to turn a relay on and off using an SPST switch or any other sensor. It can be a momentary normally open (NO) or normally closed (NC) switch if this is needed.

Figure 5.6b.   Simple logic functions can be implemented using sensors or switches and a relay as shown in this case. Here, we have a two-input AND gate using a relay. The relay will close the contacts only if the two switches (S1 and S2) are closed.

Figure 5.6c.   An OR gate is implemented with the relay as shown here. The relay will close the contacts if S1 *or* S2 is closed.

Figure 5.6d.   A two-way circuit can be implemented using SPDT switches or sensors as shown here. The relay is driven when S1 and S2 are both in condition A or both in B. In other conditions, the relay is off.

Figure 5.6e.   This circuit shows how an SPDT switch or sensor can be used to drive two relays.

Figure 5.6f.   This configuration allows the control of two relays using two wires. When S1 is in condition A, D1 is forward biased, and K1 closes the contacts. When S1 is in condition B, D2 is now forward biased, and the relay K2 is energized and closes its contacts.

Figure 5.6g.   Here we have a simple circuit to drive a DC relay from an AC source. The diode is a 1N4002 for voltages up to 50 V and currents up to 1 A. The capacitor must be in the range between 1 and 100 µF.

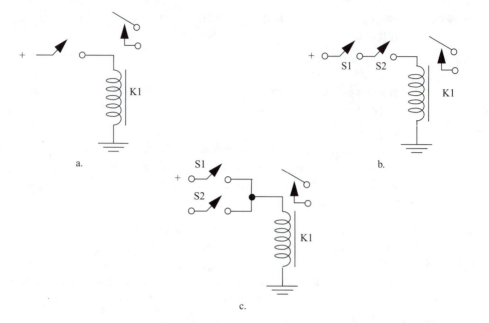

a.

b.

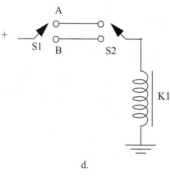

c.

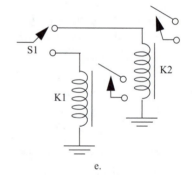

d.

e.

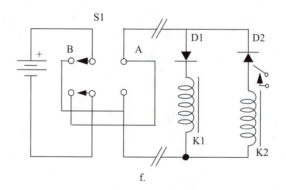

f.

Figure 5.6   *(continues)*

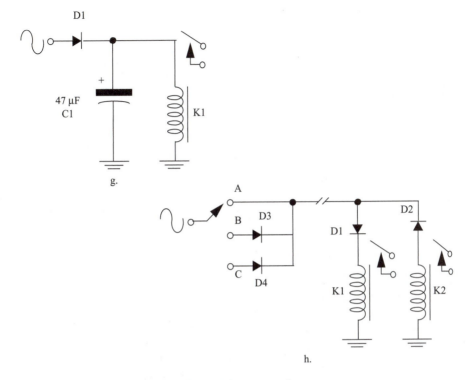

Figure 5.6 *(continued)*

Figure 5.6h. This circuit is used to control two relays using a two-wire line. When S1 is in condition A, the AC voltage is applied to D1 and D2. D1 is forward biased with the positive half-cycles, and D2 with the negative. The relays are ON. When K1 is in condition B, D3 and D1 are forward biased with the positive half cycles, and K1 closes the contacts. Changing now S1 to condition C, D4 and D2 are forward biased in the negative half cycles of the AC voltage, driving K2.

DRIVE CIRCUITS USING TRANSISTORS

Transistors are ideal elements to drive relays. With transistor amplifier stages, weak signals can be increased to levels necessary to drive a relay.

There are many ways to use transistors as relay drivers, as Fig. 5.7 shows. In this figure, we show basic configurations using common bipolar and Darlington transistors. The characteristics of each configuration are explained below.

Figure 5.7a. This is the simplest configuration that can be used to drive small, sensitive relays (50 mA or less) in the voltage range between 5 and 24 V. General-purpose transistors (such as the ones suggested in the table at the end of this section) can be used. This circuit can be driven from

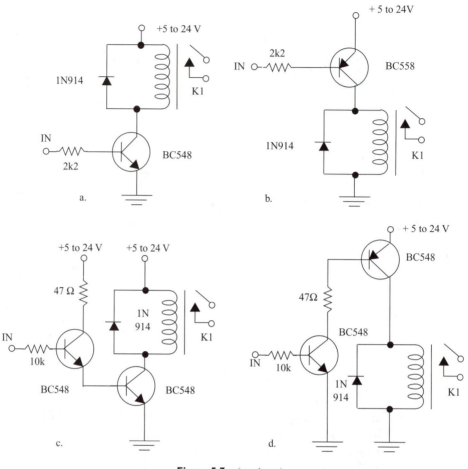

Figure 5.7   *(continues)*

TTL or CMOS signals, and it turns on the relay with positive voltages (at the HI logic level). This circuit can be driven directly from the signal provided by the parallel port of a PC. If relays with coil currents up to 500 mA must be driven, the resistor is reduced to 1 kΩ, and the transistor is replaced with any medium-power NPN such as the BD135. In this case, the signal from the parallel port is not sufficient to drive the medium-power transistor.

Figure 5.7b.   An equivalent circuit to the one shown in Fig. 5.7a, but with the use of a PNP transistor, is shown here. This circuit turns on the relay with a negative voltage, or the LO logic level if it is used with CMOS or TTL logic. Relays with coil currents up to 50 mA and voltages between 5 and 24 V can be driven. If the power needed to drive the relay is up to 500 mA, reduce the resistor to 1 kΩ and replace the

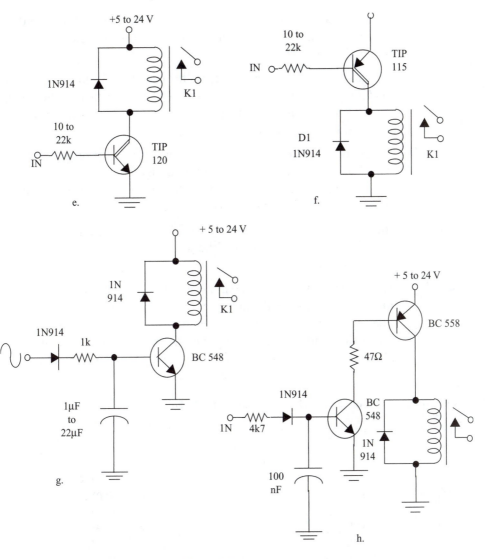

Figure 5.7 *(continued)*

transistor with a medium-power device such as the BD136 (see table at the end of the section). In this case, the signal provided by the parallel port of a PC is not sufficient to drive the circuit.

Figure 5.7c. If very low-power signals are used to drive the circuit, such as the signals supplied by sensors or low-gain stages, a Darlington configuration may be used. The configuration shown here can drive relays up to 50 mA. If higher coil currents are needed, replace transistor Q2 with a BD135 or TIP31. This circuit turns on the relay when a positive voltage or HI logic level is applied to the input. The circuit is TTL and CMOS compatible. Low-power operational amplifiers can also be used to drive this stage.

Figure 5.7d. This is another high-gain circuit, but the use of PNP and NPN transistors (complementary) is shown here. The characteristics are the same, and the PNP transistor can be replaced by a BD135 if more power is need to drive the relay (up to 500 mA).

Figure 5.7e. Darlington transistors can be used to drive high-power relays (coils up to 1 A) as shown here. Here, we see a configuration using an NPN Darlington transistor. See the table at the end of the section for information about suitable transistors.

Figure 5.7f. The PNP equivalent circuit using a Darlington transistor is shown here. See the table at the end of the section for suggestions about suitable transistors.

Figure 5.7g. AC signals (up to 100 kHz) can be used to drive a DC relay using a transistor. Tones from the output of an amplifier or decoder of a remote control receiver are example of sources for these signals. The circuit shown in the figure can drive a 50 mA relay from signals of 200 mVpp. This circuit can drive a relay from weak AC signals, since it is more sensitive than the previous ones.

Transistors Suitable for Use as High-Current Relay Drives

Type	Voltage (Vce)	Current (Ie)	Gain (hFE)	Remarks
BC547	45 V	200 mA	110–800	NPN
BC557	45 V	200 mA	75–800	PNP
BC548	30 V	200 mA	110–800	NPN
BC558	30 V	200 mA	75–800	PNP
BC549	30 V	200 mA	200–800	NPN
BC559	30 V	200 mA	125–800	PNP
BD135	45 V	1.5 A	40–250	NPN
BD136	45 V	1.5 A	40–250	PNP
1BD137	60 V	1.5 A	40–250	NPN
BD138	60 V	1.5 A	40–250	PNP
BD139	80 V	1.5 A	40–250	NPN
BD140	80 V	1.5 A	40–250	PNP
TIP31	40 V	3 A	10–50	NPN
TIP32	40 V	3 A	10–50	PNP
TIP41	40 V	6 A	15–75	NPN
TIP42	40 V	6 A	15–75	PNP
BD331	60 V	6 A	750	NPN Darlington
BD332	60 V	6 A	750	PNP Darlington
TIP110	60 V	1.25 A	500	NPN Darlington

Transistors Suitable for Use as High-Current Relay Drives *(continued)*

Type	Voltage (Vce)	Current (Ie)	Gain (hFE)	Remarks
TIP115	60 V	1.25 A	500	PNP Darlington
TIP111	80 V	1.25 A	500	NPN Darlington
TIP116	80 V	1.25 A	500	PNP darlington
TIP112	100 V	1.25 A	500	NPN Darlington
TIP117	100 V	1.25 A	500	PNP Darlington
IRF540	100 V	27 A	–	Power MOSFET

DRIVE CIRCUITS USING OTHER SEMICONDUCTORS

SCRs, power MOSFETs, isolated-gate bipolar transistors (IGBTs), and other semiconductor devices can be used to drive a relay. Figure 5.8 shows some basic configurations in which these devices are used. The main characteristics of each configuration are explained in the following text.

Figure 5.8a. Here, we show how a sensitive SCR, such as one from the 106 series, can be used to drive a relay. Relays with voltages in the range from 6 to 200 V can be used, and the coil current is limited to 2 A. If relays with more than 500 mA of coil current are driven, the SCR must be mounted on a heatsink. This circuit, when powered from a DC supply, *latches* when a positive signal (1.8 to 2 V, 200 μA) is applied to the gate of the SCR. So, even after the trigger signal is removed, the SCR remains on. To turn the circuit off, it is necessary to press S1 or to cut the power supply current momentarily.

Figure 5.8b. This circuit shows a bistable configuration driving one or two relays using two SCRs and powered from a DC power supply. The value of C (between 1 and 22 μF) is chosen experimentally as a function of the characteristics of the relays. The circuit changes state when positive pulses are applied to the gates of the SCRs.

Figure 5.8c. Here, we are driving an SCR with a transistor. This circuit is sensitive enough to be driven from TTL, CMOS logic, and even signals from the parallel port of a PC. This circuits needs only few microamperes of a positive voltage to be driven. The SCR also latches when triggered from a short pulse at the input. To turn it off, S1 must be pressed or the supply cut momentarily.

Figure 5.8d. The equivalent circuit using a PNP transistor to drive the SCR is shown here. This circuit needs only few microamperes of negative pulse to be triggered. To turn it off, we adopt the same procedure described for Fig. 5.8c.

Figure 5.8e. This circuit shows how a power MOSFET can be used to drive a relay. A positive voltage of about 2 V is need to turn on the relay. This cir-

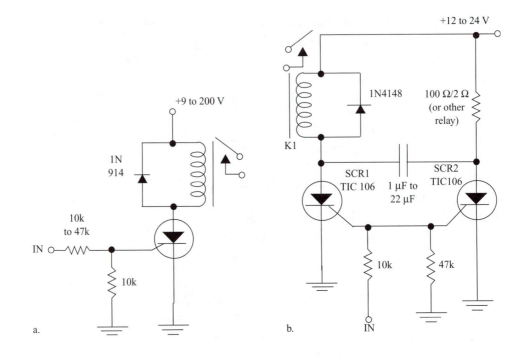

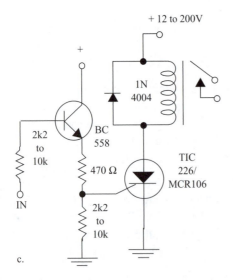

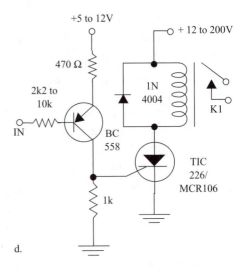

Figure 5.8   *(continues)*

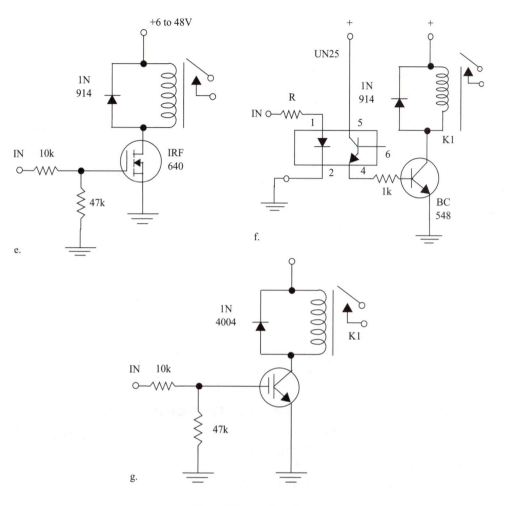

Figure 5.8 *(continued)*

cuit has a very high input impedance and can be driven directly from CMOS logic. Relays with voltage coils up to 48 V can be controlled by this configuration.

Figure 5.8f. Double isolation is provided by this circuit, where an optocoupler is used to drive a relay. This circuit is ideal for applications where the parallel port is used to drive the relay.

Figure 5.8g. Here, we have a configuration using an IGBT to drive a relay. IGBTs have the input characteristics of power MOSFETs but the control characteristics of bipolar transistors.

Obs.: Chapter 13 describes how to drive relays and other loads from the signals of a PC's parallel port.

RELAY CHARACTERISTICS

Table 5.1 shows some characteristics of typical relays.

Table 5.1   Characteristics of Common Relays

Type	Coil voltage (V)	Coil current (mA)	Coil resistance (Ω)
Reed—SPST	3	10	300
Reed—SPST	6	5	1,200
Reed—SPST	12	5	2,400
Mini/DIP	5	100	50
Mini/DIP	5	50	100
Mini/DIP	6	100	60
Mini/DIP	6	50	120
Mini/DIP	9	18	500
Mini/DIP	12	100	120
Mini/DIP	12	50	240
Mini/DIP[*]	12	10	1200

[*] Ultrasensitive

Figure 5.9 shows the pinouts of some popular relays.

TESTING AND ASCERTAINING RELAY CHARACTERISTICS

Two important characteristics of a relay must be determined if we need to use a device that is not marked with an indication of its type or origin (e.g., units obtained from surplus dealers, etc). We can actually figure out the main characteristics of a relay with the aid of one or two common multimeters, as described below.

a. *Contact terminals.*   The contacts of a relay have a near-zero ohmic resistance when closed and infinite resistance when open. Using a multimeter in any resistance (ohms) scale, we can identify the contacts that are common (C), normally open (NO), and normally closed (NC) as shown in Fig. 5.10.

   If this procedure allows us to accurately identify the common terminal, we may power up the coil of the relay to turn it on and make new measurements. The new conditions of measurement are as shown in Fig. 5.11.

b. *Triggering characteristics.*   To determine the triggering characteristics of a relay, we need a power supply with an output that is at least sufficient to drive the coil. Figure 5.12 shows a simple circuit that can be used to test relays with voltage coils up to 24 V.

   Multimeter M1 is adjusted to measure DC voltages (VDC), and multimeter M2 is adjusted to measure DC currents (DCmA up to 500 mA or more).

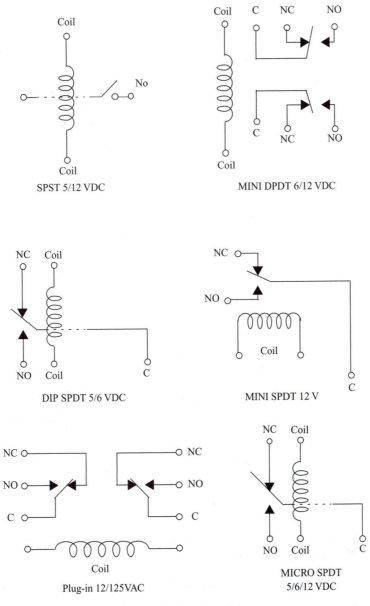

Figure 5.9

Procedure

1. Adjust P1 to 0 V at M1.

2. Open P1 slowly to raise the voltage across the relay, observing the load connected to the relay contacts.

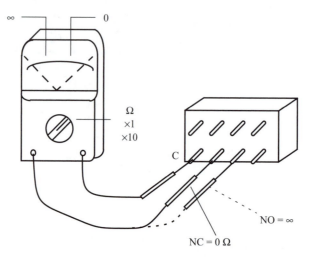

Figure 5.10

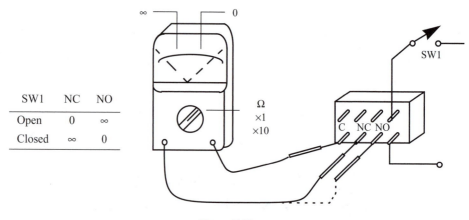

Figure 5.11

3. When the load turns on, you can read the trigger voltage of the relay on M1, and the trigger current on M2.

4. Working again with P1, reduce the voltage until the relay turns off the load.

5. At this point, you can read the hold voltage in M1 and the hold current in M2.

Note that the small difference in the trigger voltage and the hold voltage is an important characteristic to be considered when using relays. This hysteresis is your guarantee that the contacts will not bounce when the relay turns on and off.

Obs.: In some cases, is not necessary to connect a load to the contacts of the relay under test. The "click" produced by the relay when turning on and off may be enough to allow you to verify its operation.

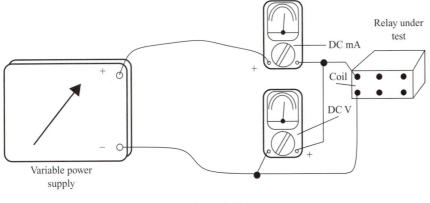

Figure 5.12

ADDITIONAL INFORMATION

a. *Connecting the contacts to control high-power loads.*   If you need to control a load that is rated for currents above the limit of the contacts of your relay, and the relay has more than one pair of contacts (e.g., DPDT), the contacts can be wired in parallel as shown in Fig. 5.13.

Some manufacturers recommend that you wire the contacts in parallel, based on the idea that is more important to increase the gap between the contacts than to increase the effective surface when controlling high currents.

b. *Debouncing.*   Problems can arise when inductive or capacitive loads are controlled and operation is disrupted by contact bounce. Some circuits, as described in Chapter 9, can be used to solve such problems.

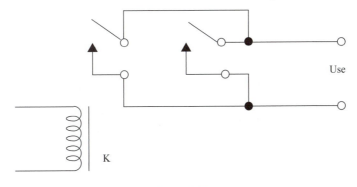

Figure 5.13

Solenoids

THE SOLENOID

Uniform magnetic fields can be generated by a long, straight coil of copper wire as shown in Fig. 6.1. If a piece of iron is placed near the solenoid, it is pulled by the magnetic forces generated by the coil, producing mechanical force. This is the operating principle of the solenoid.

The force pulling the core depends on several factors.

    a. The current passing through the coil

    b. The number of turns in the coil

    c. The material of the core

    d. The stroke (distance the core moves)

Common solenoids powered from 3 to 48 VDC supplies have currents in the range of 50 mA to 2 A. Some solenoids can release several kilograms of force when activated. Solenoids can also be powered from AC supplies.

TYPES OF SOLENOIDS

Figure 6.2 shows some of the main types of solenoids, as listed below.

    a. Linear laminated, pull

    b. Linear laminated, push

    c. Linear tubular, push

    d. Linear tubular, pull

    e. Rotary

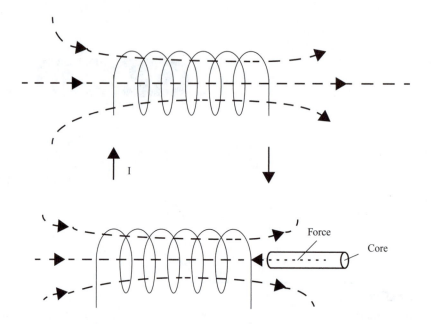

Figure 6.1

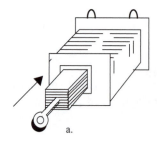

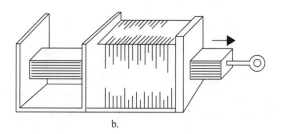

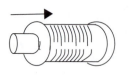

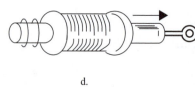

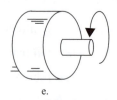

Figure 6.2

USING THE SOLENOID

Solenoids are used when only a short push or pull is needed, moving the core a short distance as shown in Fig. 6.3. You simply have to connect the solenoid to the piece to be moved, keeping in mind the required force and displacement of the core.

To meet the requirements of various practical applications for solenoids, they are available in different shapes and sizes.

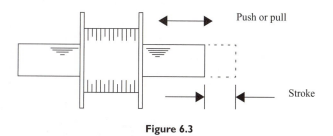

Figure 6.3

Figure 6.4 shows some applications for the solenoids in mechatronics as follows:

a. Push action

b. Pull action

c. Using a lever

d. Using a linear solenoid to obtain rotary motion

e. Rotary solenoid

f. Double action

g. Projecting objects

CHARACTERISTICS

The main specifications of a solenoid are as described below.

a. *The force/stroke created.* The force created by the plunger of a solenoid is a function of the *applied voltage* (and the current flowing through the device). The *nominal force* is the force generated when the nominal voltage is applied. This force is normally expressed in newtons (N) for linear solenoids and in terms of torque expressed in newton–meters (Nm) for rotary solenoids. Other factors that influence the solenoid force/stroke relationship are temperature and duty cycle.

b. *The nominal voltage (V_n).* The voltage that may be applied to the coil of a solenoid to produce the nominal force is the *nominal voltage* or *operating voltage.* Common types can have nominal voltage specifications between 3 and 48 V.

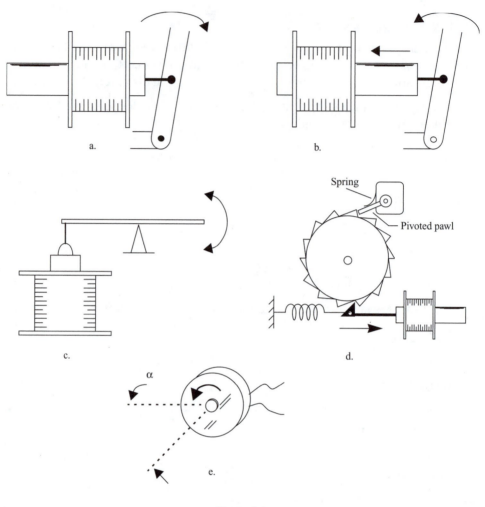

Figure 6.4

c. *Type of voltage.*   Solenoids can be powered from AC or DC sources. AC solenoids have a different construction, since they must include resources to damp
the vibrations caused by the changes in the direction of the current flowing
through the coil winding.

d. *The nominal current (I_n).*   This refers to the current that flows through the
coil when the nominal voltage is applied. This current can vary between
100 mA and several amperes for common solenoids.

e. *The nominal resistance (R_n).*   Nominal resistance is the resistance of the coil
winding. This resistance, along with the applied voltage, will determine the
nominal current of a solenoid. For common types, this resistance varies from a
fraction of an ohm to several hundred ohms.

f. *Maximum dissipation power (P_n).* Heat is generated in the coil of a solenoid when current flows through it. If a solenoid overheats, it can burn out. An important specification of the solenoid is its *maximum dissipation power.* This power can range from several milliwatts to several watts. Below, we show how to calculate this power when it is not provided by the manufacturer.

Other characteristics of importance for some applications are the mechanical life and the force/stroke curves.

Formulas:

The nominal voltage, nominal current, and the nominal resistances are interdependent. If any two of them are known, the third can be calculated using Ohm's law. The following are useful formulas for calculating these values.

Formulas:

a. Determining the resistance when the voltage and current are known:

$$R = V/I \tag{6.1}$$

b. Determining the current when voltage and resistance are known:

$$I = V/R \tag{6.2}$$

c. Determining the voltage when the current and resistance are known:

$$V = R \times I \tag{6.3}$$

where R = coil resistance (Ω)
   V = nominal voltage (V)
   I = nominal current (A)

Formula:

a. *Continuous operation.* The power dissipated by the coil of a solenoid when powered with the nominal voltage is calculated using Eq. (6.4).

$$P = V \times I = R \times I^2 = V^2/R \tag{6.4}$$

where P = power dissipated by the solenoid (W)
   R = coil resistance (Ω)
   V = nominal voltage (V)
   I = nominal current (A)

b. *Pulsed operation* (considering the duty cycle).

$$P = V \times I \times DT/100 \tag{6.5}$$

where P = power dissipated by the solenoid (W)
  R = coil resistance (Ω)
  V = nominal voltage (V)
  I = nominal current (A)
  DT = duty cycle (%)

In any application, P must be lower than P_d (the maximum dissipation power) to avoid overheating the solenoid.

Obs.: Since, in many applications, the solenoid is activated for only a short time, the maximum dissipation power is not necessarily an absolute limit.

DUTY CYCLE

The duty cycle of a solenoid is the ratio of the time on to the total operational cycle time. For optimal operation, the duty cycle should be minimized.

Formula:
The duty cycle can be expressed as

$$DT = 100 \times \frac{T_{on}}{T_{on} + T_{off}} \tag{6.6}$$

where DT = duty cycle in percent
  T_{on} = time the solenoid is on
  T_{off} = time the solenoid is off

Obs.: The duty cycle also determines the amount of power dissipated by the solenoid in a particular application.

DETERMINING THE CHARACTERISTICS OF A SOLENOID

Solenoids can be assembled from common parts such as cardboard tubes, enameled wire, and a screw used as the core (see Fig. 6.5.) You can calculate the number of turns needed for an application but, in practice, there will be differences between the theoretical calculations and the characteristics of the actual device.

A question now arises: How can we determine the characteristics of a home-made solenoid, or of a device that we have salvaged from a surplus parts bin or a dis-

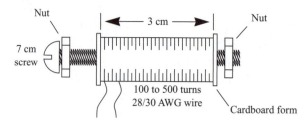

Figure 6.5

carded appliance, when we have no information about it? The methods are described below.

Voltage

This is not an easy attribute to determine. If we know nothing about the operating voltage of the solenoid, we have to assume that it can be used for our application as long as it produces enough force without overheating. In this case, the simplest way to test the device is to connect it to a variable power supply as shown in Fig. 6.6.

While raising the voltage, we can measure the force produced and determine whether it is sufficient for the application we have in mind. At the same time, we monitor the temperature of the coil. For a safe operation, the maximum applied voltage should not heat the coil to more than 50°C. A force × voltage curve, as shown in Fig. 6.7, can be created from this experiment.

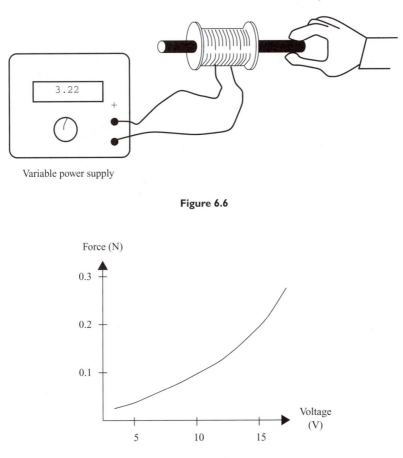

Variable power supply

Figure 6.6

Force (N)

Voltage (V)

Figure 6.7

Current

The current in a DC-powered solenoid depends on the applied voltage. Normally, a current rating is provided by the manufacturer. But if you have to determine the current rating yourself, you can use the circuit shown in Fig. 6.8.

The current through the coil can be read directly from the scale of the multimeter. If the resistance and the nominal coil voltage of a solenoid are known, the current can be calculated using Ohm's law.

Formula:

$$I = V/R \tag{6.7}$$

where I = nominal current of the solenoid, or the current under certain voltage (A) in a specified application
V = voltage applied to the coil (V)
R = nominal resistance of the coil (Ω)

Force

The force generated by a solenoid can be measured using a dynamometer. Figure 6.9 shows three suitable processes for determining the force created by a solenoid: (a) using a dynamometer, (b) using a common balance scale, and (c) using known weights.

It is important to observe that, when using a spring dynamometer, the lever is needed to increase the displacement, since the indication of force depends on this displacement.

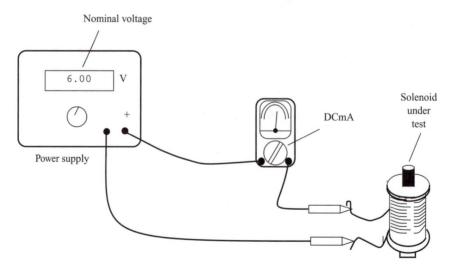

Figure 6.8

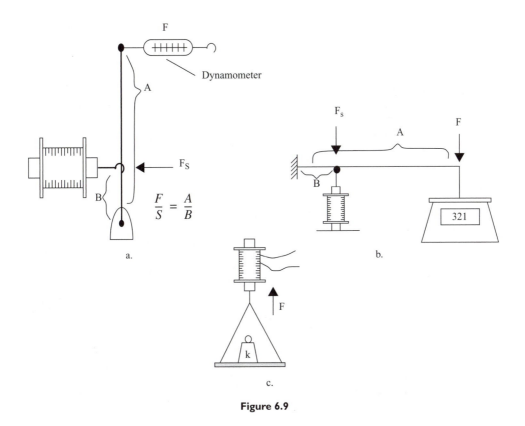

Figure 6.9

DRIVING A SOLENOID

The simplest way to drive a solenoid is by connecting it to a power supply that produces the nominal voltage, as shown in Fig. 6.10. But, in more complex projects, electronic circuits can be used to drive the solenoid, using weak signals such as the outputs from the parallel port of a PC, logic circuits, microprocessors and microcontrollers, and low-power sensors. Figure 6.11 shows some configurations that can be used to drive solenoids, as listed below.

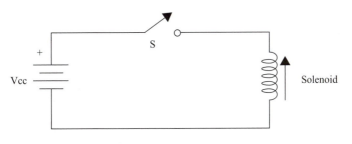

Figure 6.10

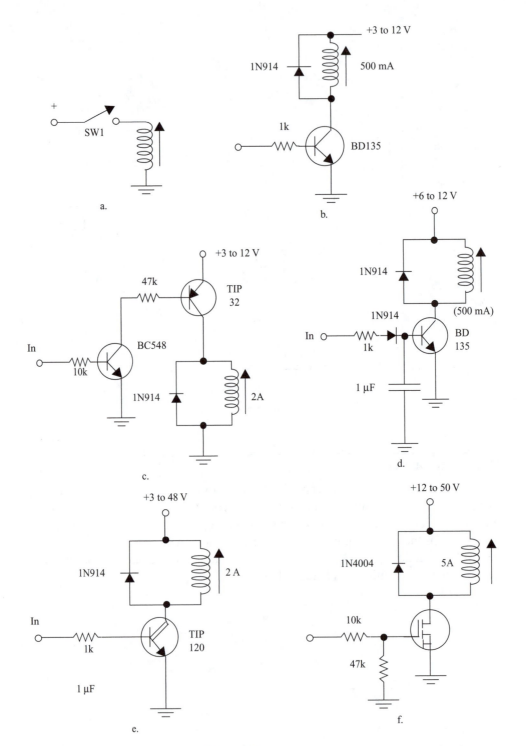

Figure 6.11 *(continues)*

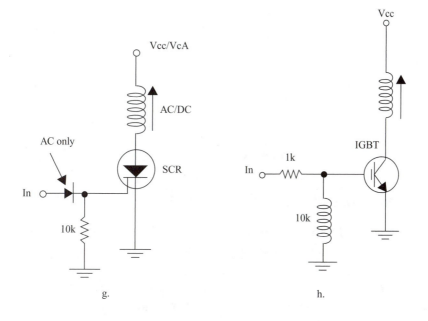

g.

h.

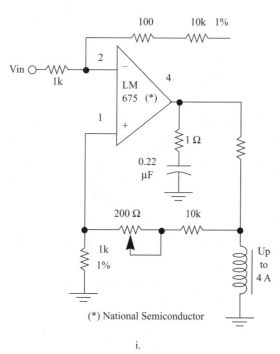

(*) National Semiconductor

i.

Figure 6.11 *(continued)*

Figure 6.11a.   Direct driver

Figure 6.11b.   Bipolar transistor

Figure 6.11c.   Two-stage bipolar transistor

Figure 6.11d.   AC driver

Figure 6.11e.   Darlington transistor

Figure 6.11f.   Power MOSFET

Figure 6.11g.   Silicon controlled rectifier (SCR)

Figure 6.11h.   Isolated-gate bipolar transistor (IGBT)

Figure 6.11i.   Dedicated IC

See Chapters 4 and 5 for additional configurations. The same circuits used to drive DC motors and relays can be used to drive solenoids.

DEFINITIONS

Some important solenoid terminology is provided below.

Stroke	Stroke is the total expected movement of the solenoid plunger when power is applied. It also can be defined as the gap between the moveable plunger and the fixed pole piece when the solenoid is off.
Force	This refers to the maximum load that the solenoid can push or pull when energized.
Duty cycle	Duty cycle is the ratio between the time the solenoid is energized and the total time of a full operating cycle [see Eq. (6.6)].
Heat rise	This is the stabilized increase in coil wire temperature that occurs during operation at a rated voltage and duty cycle in the device's intended ambient conditions.
Return springs	These springs are used to return the plunger when the coil current is turned off.
Potted coil	For environmental protection, the coil may be hermetically sealed (potted).
Response time	This is the time delay between the instant the power is switched on and the moment the plunger reaches the end of its stroke. This time falls in the range of a few milliseconds (5 to 250 ms typically, depending on the size and stroke of the solenoid).

CHAPTER 7

Servomotors

THE SERVOMOTOR

A servomotor (commonly referred to simply as a *servo*) is a motor whose output shaft can be moved to a specific angular position by sending it a coded signal. As long as the coded signal is applied to the input of the servo, it will maintain the position of the shaft. If we change the coded signal, the angular position of the shaft also changes. Figure 7.1 shows a typical servo for remote control (RC) applications.

The most popular use of the servo is in radio-controlled airplanes to position control surfaces such as elevators and rudders. But, of course, they are a good choice for other mechatronic applications such as cars, robots, robotic arms, and so on. High-power units can be used in other applications such as industrial automation.

OPERATING PRINCIPLE

A servomotor is a closed-loop device. When sent a signal, it adjusts itself to match the signal. If the signal changes, the motor readjusts to match it.

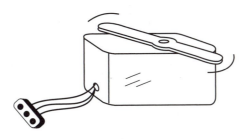

Figure 7.1

Servo motors are geared DC motors with the closed-loop circuitry incorporated within them.

BASIC PRINCIPLE

The basic configuration of a servomotor is shown in Fig. 7.2. A DC motor is used to move a gearbox with a large reduction ratio.

The final shaft, moving at a very low speed, imposes a force on the external load and simultaneously acts on the axis of a feedback potentiometer. This potentiometer senses the position of the axis and sends a corresponding voltage to an operational amplifier (connected as a voltage comparator).

This voltage is compared to the input voltage (which determines the desired position of the shaft), producing a voltage in the output of the comparator. This voltage powers the motor such that the shaft moves in the necessary direction to align with the angle that corresponds to the voltage applied to the input.

STANDARD CONFIGURATION

Another way to control the position of the shaft of a servo is through the use of a coded signal. This method is common in standard servos used with RC applications such as cars, airplanes, toys, and other devices.

As shown in Fig. 7.3, the signal applied to the control input of the servo is a train of pulses with a total period of 18 ms. The duration of the pulse will determine the position of the shaft in a range of 180° as shown in the figure.

If the pulse is 1.5 ms long, the servo motor will be positioned in the middle range. If the pulse is 1 ms long, the servo will move 90° to the left, and if the pulse is 2 ms

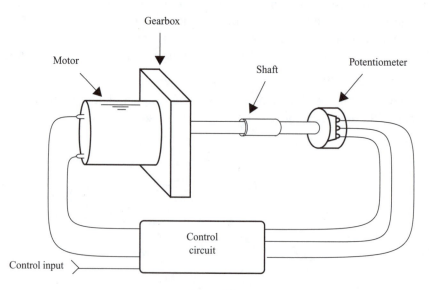

Figure 7.2

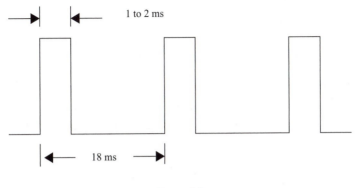

1 to 2 ms

18 ms

Figure 7.3

long (corresponding to a frequency of approximately 50 Hz), the servo will move 90° to the right.

Other types might have a pulse length range between 1.25 and 1.75 ms. In this case, the shaft will be at center with 1.50 ms, 90° to the left with 1.25 ms, and 90° to the right with 1.75 ms.

TYPES

The standard servo has three wires: one for power (between 4 and 6 V), one for ground, and the last for control. The colors of the wires are generally (1) red = +VDC, (2) black = ground, and (3) white = control.

The size and shape of a servomotor depends on the intended application. Figure 7.4 shows two common types of servo motors used in RC and mechatronics. They are low-power servos that can be powered from single cells, small batteries, and other DC supplies in the range of 100 mA to 2 A. There are also high-power types that are powered from AC supplies and used in industrial and mobile applications.

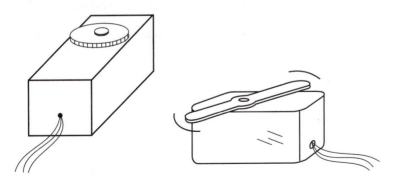

Figure 7.4

CHARACTERISTICS

As with other types of motors, servos have important electrical and mechanical characteristics. The main ones are described below.

Power Supply Voltage

The servo types used in RC normally can be powered from supplies in the range of 4 to 6 V. Special types can be found that operate at voltages outside this range.

Current

Here we refer to the current drawn by the motor when the maximum torque is applied to an external load. As with standard DC motors, this current varies with the load. Common servos have maximum currents in the range of 100 mA to 2 A.

Operating Speed

This is defined as the time required for the shaft to reach a specified position (normally 60°). Common servos have operating speeds in the range of 0.05 to 0.2 s/60°.

Torque

As with other types of motors, the torque is given in kg/cm or N-cm. Typical values are in the range of 0.5 to 10 kg/cm. See Chapters 4 and 6 for information about how to measure the torque of a motor or a solenoid. The procedures described therein can be applied to servo motors as well.

Control Pulse

This refers to the type of pulse used to position the shaft. Two main types are used in RC applications: center position in 1–2 ms and 1.25–1.75 ms.

In the 1–2 ms type, the pulse width ranges from 1.25 and 1.75. In the 1.25–1.75 ms type, the range is between 1 and 2 ms. Tables 7.1 and 7.2 show pulse duration vs. position angle for both types of servos.

Table 7.1   Pulse Duration × Position Angle (for 1–2 ms Servos)

Duration (ms)	Angle (degrees)	Duration (ms)	Angle (degrees)
1	−90	1.6	18
1.1	−72	1.7	36
1.2	−54	1.75	45
1.25	−45	1.8	54
1.3	−36	1.9	72
1.4	−18	2	90
1.5	0		

Table 7.2 Pulse Duration × Position Angle (1.25–1.75 ms Servos)

Duration (ms)	Angle (degrees)	Duration (ms)	Angle (degrees)
1.25	−90	1.55	18
1.30	−72	1.60	36
1.35	−54	1.65	54
1.40	−36	1.70	72
1.45	−18	1.75	90
1.50	0		

Resolution

Resolution defines the precision with which the shaft is positioned when it receives an external command signal. Typical servos have resolutions in the range from 1° to 10°.

Size and Weight

This information is important in mechanical project design. Small RC servos have a typical weight range between 15 and 200 g.

Formulas:

a. Pulse duration × position

$$t = \frac{(\alpha - 90) \times (t_2 - t_1)}{180} + t_1 \tag{7.1}$$

b. Position × pulse duration

$$\alpha = \frac{(t - t_1) \times 180}{t_2 - t_1} + 90 \tag{7.2}$$

where t = pulse duration for a particular position (ms)
α = shaft position angle (degrees)
t_2 = upper limit of the pulse length (ms), 1.75 or 2
t_1 = lower limit of the pulse length (ms), 1 or 1.25

BUILDING A SIMPLE SERVO

You can build your own experimental servo using common parts. Figure 7.5 shows a simple project. This servo is controlled by voltage rather than pulses (but we can adapt it to operate with pulses). The control circuit is shown in Fig. 7.6. In Fig. 7.5,

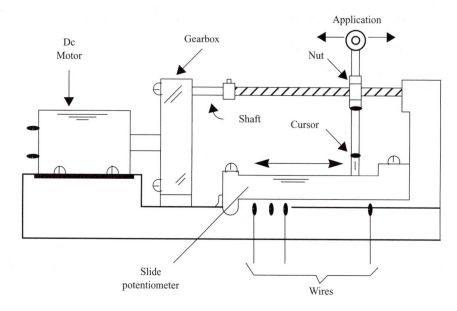

Figure 7.5

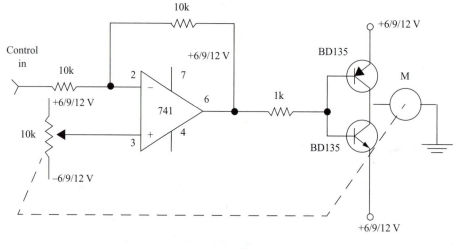

Figure 7.6

a gearbox is attached to a worm screw that moves a nut. The nut is coupled to a slide potentiometer that provides electrical feedback to the circuit.

When you apply a voltage to the input of the circuit (signal), the motor moves until the potentiometer reaches the position at which the feedback voltage equals the input voltage (zero output). If you change the input voltage, the motor is powered until the potentiometer reaches the new position at which the output is zeroed.

ADVANTAGES OF SERVO SYSTEMS OVER STEPPER SYSTEMS

There is one basic difference between servo and stepper control systems: the use of feedback. Servos have a position encoder attached to the drive motor that reports the actual position of the motor shaft back to the controller. If any positioning errors exist, the servo controller may take corrective action to ensure that the motor moves to the proper position.

In comparison, stepper controllers can only issue a move command, and you must simply hope that the motor is capable of following it. You have no way of being sure that the motor has indeed reached the desired position. The presence of feedback in a servo controller system results in several benefits, enumerated below.

1. *No lost pulses.*   Servo systems know exactly where the motor is at all times, so all step commands are executed.

2. *Full torque at high speeds.*   Stepper motor torque falls off as motor speed increases, due to electrical time constants and poor current utilization. Servo motors do not have this problem and may generate full torque at high rpm.

3. *Quiet, smooth operation.*   Servo systems are inherently smooth due to their high encoder resolutions—typically at least ten times finer than a stepper motor's number of positions per revolution.

4. *Zero holding current.*   Stepper motors must have high currents applied at all times, even while they are stationary, with little or no load. Servo systems draw power only as required, and the power drain is proportional to the load torque applied to the motor.

5. *True microstepping.*   Stepper motor resolution may be increased via a process, called *microstepping,* in which currents are applied to the motor windings in proportion to the desired position between normal steps.

CONTROL CIRCUITS

Servo motors are powered from DC supplies and controlled by a pulse source. Special circuits can be used to convert information that has been picked up from sensors, generated by external control pads or microcontrollers, or supplied by the parallel port of a PC or other computer.

Figure 7.7 shows three circuits that are suitable for the control of servomotors. The circuit shown in Fig. 7.7a uses two 555 ICs to generate the control pulses. One of the ICs produces the delay, and the other generates the pulse width. The delay (complete cycle) must be adjusted to 18 ms. The other circuit must be adjusted according to the position of the servo. A potentiometer or other resistive sensor can be used in such cases. The system can be calibrated by wiring an oscilloscope to the output of each IC (pin 3).

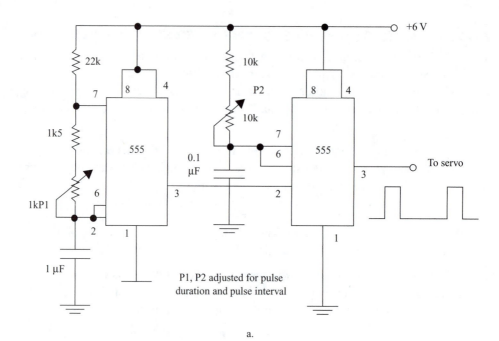

P1, P2 adjusted for pulse
duration and pulse interval

a.

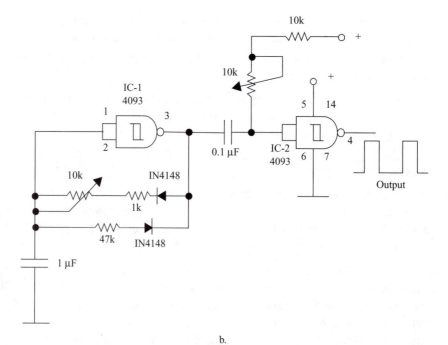

b.

Figure 7.7   *(continues)*

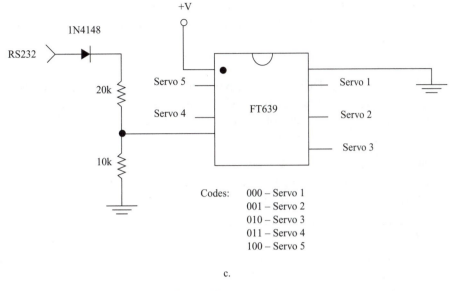

Figure 7.7 *(continued)*

The circuit shown in Fig. 7.7b uses two gates of a CMOS circuit. The timing of the command cycle is produced by the oscillator formed by IC-1. The trimmer potentiometer must be adjusted so that the circuit produces 18 ms pulses. This circuit commands IC-2, a monostable with a time constant RC in the range between 1 and 2 ms. P1 is the control element that adjusts this time constant.

The components of the circuit must be chosen to find the correct timing and frequencies by compensating for their tolerances. An oscilloscope can be used to make these adjustments.

In Fig. 7.7c, a dedicated IC manufactured by FerretTronics (www.ferrettronics.com) is used—the FT639. The FT639 is a servo controller chip. It will govern up to five radio-controlled servos through one 2400-baud serial line. No components are needed for a 0.5 V serial line such as the ones found in microcontrollers like the Parallax BASIC Stamp.

Each RC servo has 256 positions. To send the servo position to the FT639, two commands are needed. The first contains the servo number and lower four bits—the positional number. The second command contains the upper four bits of the positional number. Additional information about the codes and the operational modes can be found in the web site mentioned above.

TESTING A SERVO MOTOR

You can find RC servos in toys such as RC model cars and airplanes, and they can be adapted for application in your particular mechatronic device. To test a servo motor, at least some of its characteristics must be known, e.g., the power supply voltage or the method of control (pulse, voltage, current, etc.). The following procedures are suitable for testing common RC servos with voltages in the range of 4.8 to 6.0 V.

Electrical Testing

Power the servo from a DC supply, applying the appropriate voltage, and use the circuit of Fig. 7.7a or 7.7b to verify the shaft position. If you have an oscilloscope, you can determine the type of command required by the servo (pulse length and cycle).

Mechanical Testing

The primary mechanical test that needs to be made on a servo is torque measurement. This can be done as described in Chapter 4.

COMMERCIAL TYPES

The following are some servos that appear in the catalog of Images SI Co. (www.imagesco.com/catalog/index.html).

1. HS-50 Feather Servo
    Operating speed: 0.09 s/60° at 4.8 V
    Output torque: 0.6 kg/cm at 4.8 V
    Weight: 5.8 g

2. HS-81 Micro Servo
    Operating speed: 0.09s/60° at 6.0 V
    Output torque: 3 kg/cm at 6.0 V
    Weight: 16.6 g

3. HS-85BB+ Micro Servo
    Operating speed: 0.14 s/60° at 6.0 V
    Output torque: 3.5 kg/cm at 6.0 V
    Weight: 19.2 g

4. HS-300 Standard Servo
    Operating speed: 0.16 s/60° at 6.0 V
    Output torque: 3.5 kg/cm
    Weight: 44.5 g

5. HS-715BB Sail Servo
    Operating speed: 0.22 s/60° at 6.0 V
    Output torque: 13.5 kg/cm at 6.0 V
    Weight: 102 g

"HACKING" A SERVO

Servos are indicated for applications in which the shaft must be positioned via an external command. However, modifications can be made to convert a conventional servo into a common gearhead motor with built-in electronics. The final result depends on the type of servo to be modified, but the most common one that is suitable for this task is the Futaba S-148.

The basic idea is given in the following steps (which can vary, depending on the type and application):

1. Open the servo and remove the plastic gears.

2. Remove the circuit board.

3. Remove the potentiometer and replace it with a fixed resistor network of the same value (each resistor may have half of the nominal value of the potentiometer, thereby simulating the center position).

4. Reassemble the circuit board.

5. Modify the gear configuration by removing the mechanical stop.

6. Replace the gears.

7. Reclose the box.

CHAPTER 8

Shape Memory Alloys

INTRODUCTION TO SMAS

Shape memory alloys (SMAs) are materials that have the ability to return to a predetermined shape when heated. When an SMA is cold (i.e., below its *transformation temperature*), it has a very low yield strength and can be deformed quite easily. Below the transition temperature, it has a tendency to retain the deformed shape. In most cases, the transition temperature of the SMA is chosen such that the transformation point of the material is well above room temperature.

If the material is heated to its transformation temperature, it undergoes a change in crystalline structure. This change causes it to return to its original shape. If the SMA encounters resistance during this transformation, it can generate extremely large forces, providing a unique mechanism for remote actuation.

The most common shape memory material is an alloy of nickel and titanium called *nitinol*. This alloy has very good electrical and mechanical properties, long fatigue life, and high corrosion resistance. As an actuator, it is capable of up to 5 percent strain recovery and 50,000 psi restoration stress over many cycles.

Nitinol also has resistance properties that enable it to be actuated electrically by joule heating. When an electric current is applied to nitinol, it can generate enough heat to cause the phase transformation.

PHYSICAL PROPERTIES OF NITINOL

Nitinol is available in the form of wire, rod and bar stock, and thin film. Its physical properties are as listed below.

Density: 6.45 g/cc

Melting temperature: 1240–1310°C

Resistivity (high-temperature state): 82 μΩ-cm

Resistivity (low-temperature state): 76 μΩ-cm

Thermal conductivity: 0.1 W/cm-°C

Heat capacity: 0.077 cal/g-°C

Latent heat: 5.78 cal/g; 24.2 J/g

Magnetic susceptibility (high-temperature): 3.8 μemu/g[*]

Magnetic susceptibility (low-temperature): 2.5 μemu/g

MECHANICAL PROPERTIES OF NITINOL

Ultimate tensile strength: 754–960 MPa or 110–140 ksi

Typical elongation to fracture: 15.5 percent

Typical yield strength (high-temperature): 560 MPa, 80 ksi

Typical yield strength (low-temperature): 100 MPa, 15 ksi

Approximate elastic modulus (high-temperature): 75 GPa, 11 Mpsi

Approximate elastic modulus (low-temperature): 28 GPa, 4 Mpsi

Approximate Poisson's ratio: 0.3

ACTUATION CHARACTERISTICS

Energy conversion efficiency: 5%

Work output: ~1 joule/gram

Available transformation temperatures: –100 to +100°C

COMMON SMA CHARACTERISTICS

In Table 8.1, we display some important characteristics of SMAs (information derived from the Robot Store catalog).

Table 8.1   Characteristics of Common SMA

Wire name	Wire diameter (microns)	Linear resistance (Ω/m)	Typical current (mA)	Deform. weight[*] (g)	Recovery weight (g)	Typical rate[†] (LT/HT)[‡]
Flexinol 025	25	1770	20	2	7	55/n.a.
Flexinol 037	37	860	30	4	20	52/68

*emu = electromagnetic unit

Table 8.1 Characteristics of Common SMA *(continued)*

Wire name	Wire diameter (microns)	Linear resistance (Ω/m)	Typical current (mA)	Deform. weight[*] (g)	Recovery weight (g)	Typical rate[†] (LT/HT)[‡]
Flexinol 050	50	510	50	8	35	46/67
Flexinol 100	100	150	180	28	150	33/50
Flexinol 150	150	50	400	62	330	20/30
Flexinol 250	150	20	1000	172	930	9/13
Flexinol 300	300	13	1750	245	1250	7/9
Flexinol 375	375	8	2750	393	2000	4/5

[*] Multiply by 0.0098 for force in Newtons (N).
[†] Cycles per minute in still air at 20°C.
[‡] LT = low temperature, 70°C; HT = high temperature, 90°C.

HYSTERESIS

When the wire heats under a constant force, its contraction follows the right-hand curve shown in Fig. 8.1. When the temperature reaches the point A_s, the wire has started to shorten. Nearly complete contraction is achieved when the temperature reaches A_f.

As the wire cools, the left-hand curve is now followed, starting at lower right and passing through M_s and M_f. M_s is the point where the wire begins to relax again, and M_f is the point where the wire is almost fully relaxed.

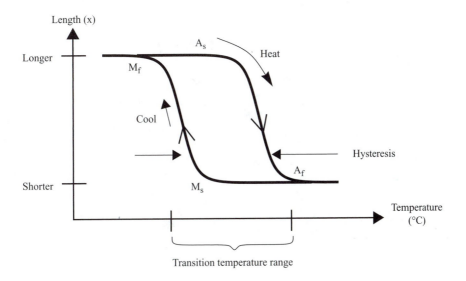

Figure 8.1

USING SHAPE MEMORY ALLOYS

Shape memory alloys can be used wherever a short pull or push is needed. One of the most popular applications for SMAs is in *robot muscles*. In the following text, we look at some important considerations in the use of SMAs, with the aim of determining whether an SMA is the best choice for a particular project.

MECHANICAL CONSIDERATIONS

Figure 8.2 gives an example of a typical SMA application in which a wire works as an *electric muscle to* drive a robot arm. When the SMA is driven by a current source, it acts on the arm to release force. The amount of force generated at the end of the arm can be calculated from the SMA force and the characteristics of the arm (see Chapter 2 for formulas and procedures).

It is important to choose the proper wire gauge for your application. If the force generated at the end of the arm exceeds the maximum capacity of the wire, the wire will break.

Another characteristic to be considered is mechanical fatigue. An SMA can contract and release a limited number of times, so life expectancy is an important factor.

ELECTRICAL CONSIDERATIONS

The circuit that you will need to drive the SMA is another consideration in any project. Excess current can break the SMA, and insufficient current will not generate enough heat to bring the wire to its transition temperature; hence, it will not contract as required. You therefore need to consult the manufacturer's specifications and use a power source that provides the correct amount of current.

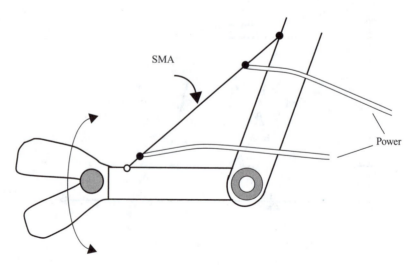

SMA

Power

Figure 8.2

Below, we consider two types of power sources: a simple voltage source and a constant-current source.

Simple Voltage Source

In simple cases, if we know the voltage of the power source and the specific resistance of the SMA (in Ω/m or Ω/cm), we can use a simple calculation to determine the length of wire necessary for a particular operating current.

First, we use Ohm's law to compute the total resistance that the wire must have to produce a flow that matches the SMA's nominal current rating so as to achieve the desired change in shape.

Formulas:

$$R = V/I \qquad (8.1)$$

where R = resistance in ohms
   V = voltage of the power source in volts
   I = current required by the SMA in amperes

In the second step, we calculate the length of SMA that produces the required resistance.[*]

$$L = R/\rho \qquad (8.2)$$

where L = final length in meters or centimeters
   R = total resistance in ohms
   ρ = resistivity in ohms/m or ohms/cm

Cells and batteries are commonly used to power SMA devices, as shown in Fig. 8.3.

For practical purposes, the applied current must be 10 to 20 percent larger than the calculated one to guarantee proper performance after compensating for some voltage drop in the source. If the voltage from the source is higher than required for the length and type of SMA used, a resistive voltage divider can be used as shown in Fig. 8.4.

Resistance R is calculated using the following formulas:

Formulas:

   a. Value of R

$$R = (V_{in} - V_{SMA})/I \qquad (8.3)$$

---

[*]In critical applications, we must also consider that the resistance of the SMA changes as it is heated; resistance increases with heat, so the current flow drops.

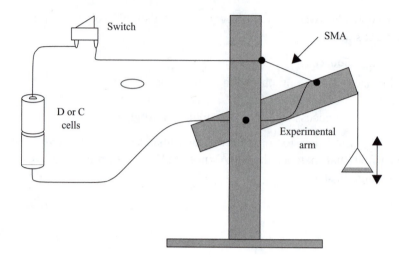

Figure 8.3

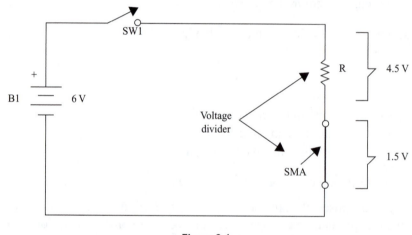

Figure 8.4

where R = added resistance in ohms
V_{in} = input voltage in ohms
V_{SMA} = voltage required to drive the SMA in the application
I = current passing through the SMA in amperes

b. The dissipation of R is given by

$$P = R \times I^2 \qquad (8.4)$$

where P = power dissipated in watts

R = resistance of the added resistor in ohms

I = circuit current in amperes

Obs.: Use a wire-wound resistor with at least twice the calculated value to avoid exceeding the device's dissipation limit.

Constant-Current Source

Although a simple current source is applicable, optimal SMA performance is obtained through the use of a constant current source. This kind circuit of circuit maintains a constant circuit current independently of the circuit's resistance and the input voltage, and it can do so within a wide range of both voltage and resistance.

Figure 8.5 shows a simple constant-current source. It uses an adjustable voltage regulator IC such as the LM150/LM317/LM350T or any equivalent. The LM350T, for instance, can provide a load with currents up to 3 A. (See Chapter 11 for more information about voltage regulator ICs.)

The value of R depends on the desired current in the load. It can be calculated using Eq. (8.5).

Formula:

$$R = 1.25/I \qquad (8.5)$$

where R = circuit resistance in ohms

I = required current to drive the SMA in amperes

In Eq. (8.5), the 1.25 value is the internal reference voltage of ICs such as the LM150, LM350, and LM317. If another type of IC is used, the specification sheet will give the internal voltage reference to be substituted for the 1.25 value.

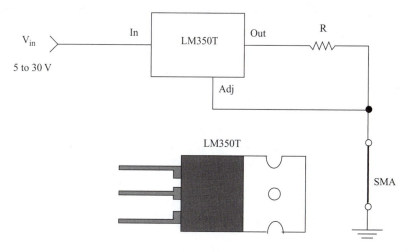

Figure 8.5

The value of V_{in} is derived from the resistance of the SMA or the voltage necessary to drive it. If we know the SMA resistance and the current passing through it, we can calculate the voltage required to drive it using Ohm's law.

Formula:

$$V = R_{SMA} \times I \qquad (8.6)$$

where V = voltage across the SMA in volts
    R_{SMA} = resistance of the SMA in ohms
    I = current in amperes

V_{in} must be at least 2 or 3 V higher than V_{SMA}. See Chapter 11 for the maximum V_{in} for a particular IC.

PRACTICAL CIRCUITS

Figure 8.6 shows some practical circuits that are suitable for power and control of SMAs. Below, we explain the operation and use of each one.

- *Figure 8.6a.*   This represents the simplest constant-current source, with values calculated for a 200 mA SMA.

- *Figure 8.6b.*   This is a constant-current source that uses a transistor. This circuit can control SMAs requiring up to 500 mA.

- *Figure 8.6c.*   This voltage source can be adjusted using P1. The voltage is adjusted to match the current needed to drive the SMA. It is not a great current source.

- *Figure 8.6d.*   This is a pulsed (pulse width modulated, PMW) source. The median current passing through the SMA can be adjusted by varying the duty cycle of the voltage output. This circuit is less sensitive to power supply voltage changes.

- *Figure 8.6e.*   Figure 8.6e shows a transistor-controlled SMA. The circuit can be turned on and off by applying a positive voltage to the base of Q1. *The transistor must be mounted on a heatsink.*

- *Figure 8.6f.*   This figure shows a timed circuit. It turns an SMA on and off at time intervals that are determined by R_a, R_b, and C, according to the following formulas.

Formulas:

$$\text{Time on: } T = 1.1 \times R_a \times C \qquad (8.7)$$

$$\text{Time off: } T = 1.1 \times (R_a + R_b) \times C \qquad (8.8)$$

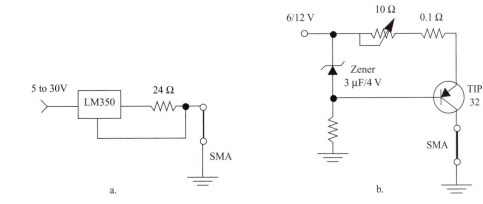

a.

b.

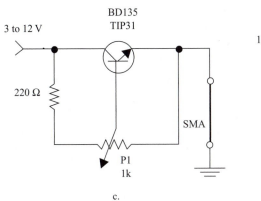

c.

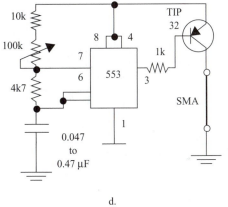

d.

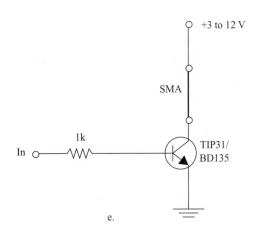

e.

Figure 8.6

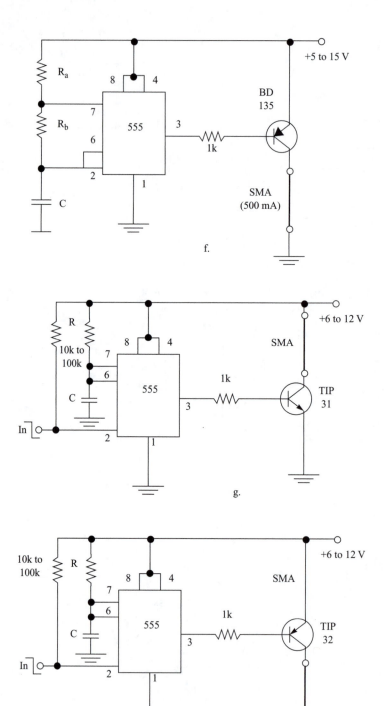

Figure 8.6

- *Figure 8.6g.* This figure shows another timed circuit. Here, monostable action drives the SMA when a negative pulse is applied to the input. The time during which the output remains on is given by

Formula:

$$T = 1.1 \times R \times C \qquad (8.9)$$

- *Figure 8.6h.* This is the same circuit as shown in Fig. 8.6f, but this one is used when the time on must be greater than the time off.

DETERMINING THE CHARACTERISTICS OF AN SMA

If you have a measured length of SMA wire, but you don't know its characteristics, you can determine them easily using the circuit arrangement shown in Fig. 8.7. The power supply must be rated to 2 A or more, depending on the SMA under test. If you have only a multimeter, first connect it at points (A) to find the transition voltage. Then connect it at points (B) to find the current.

Procedure

a. Adjust the output of the power supply (using P1) to 0 V.

b. Then, slowly open P1 to increase the output voltage while observing the SMA.

c. When you reach the transition current, the SMA will change its shape, thereby moving the test arm.

d. Increase the voltage a little bit more—no more than 10 percent above the previous value. Read the voltage at points (A).

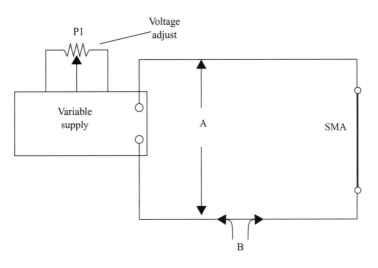

Figure 8.7

e. Read the current at points (B).

f. With these values, you can determine the following:

– The resistance of the SMA

$$R = V(A)/I(B) \qquad (8.10)$$

where R = resistance of the SMA in ohms
$V(A)$ = voltage reading in (A)
$I(A)$ = current reading in (B)

– The current to drive the SMA, $I(B)$

– The resistivity of the SMA as follows:

$$\rho = R/L \qquad (8.11)$$

where ρ = resistivity in ohm-cm
R = resistance in ohms
L = length in centimeters

MORE PRACTICAL INFORMATION FOR SMA USE

MINIMUM BEND RADIUS
If an SMA wire is bent too sharply (see Fig. 8.8), there is a risk of damage. Table 8.2 shows the minimum recommended bend radii for various wire diameters.

CONNECTION METHODS
To avoid contact problems when using the SMA wires, it is important to use some of the recommended connection methods shown in Fig. 8.9.

BIAS FORCE MECHANISMS
Figure 8.10 shows some ways to create bias force against SMA wires. Methods are as described below:

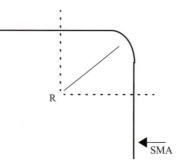

R

SMA

Figure 8.8

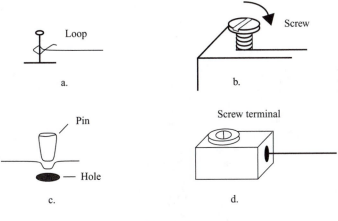

Figure 8.9

Table 8.2 Minimum Bend Radius

Wire diameter (μm)	Minimum bend radius (mm)	Wire diameter (μm)	Minimum bend radius (mm)
25	1.25	150	7.50
37	1.85	200	10.0
50	2.50	250	12.5
75	3.75	300	15.0
100	5.00	375	16.7
125	6.25		

- *Figure 8.10a.* The force is constant along the length A-B.
- *Figure 8.10b.* The force increases with the length A-B.
- *Figure 8.10c.* The force decreases with length.

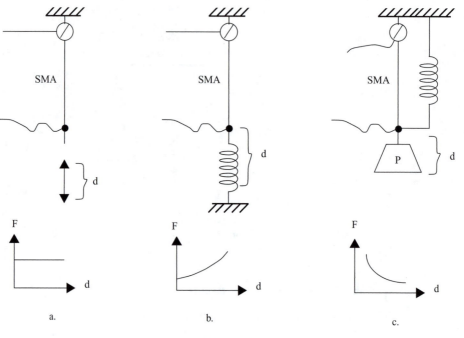

Figure 8.10

Sensors

ON-OFF SENSORS

The purpose of this chapter is provide suggestions about the use of mechanical switches, reed switches, and homemade devices as sensors. Pairs of metal blades, common on-off switches, pushbuttons, microswitches, and reed switches can be used as sensors in robots and mechatronic projects.

A motor can be reversed when a robot is stopped by an obstacle or a mechanical arm bumps up against an object. A mobile robot can change its direction if a sensor detects a wall in its path. You can add mechanical sensors to your projects using some common components and configurations as described below.

ON-OFF SWITCHES, MICROSWITCHES

On-off switches can be adapted easily for use as position sensors in many applications. Figure 9.1 shows how can we adapt a slide on-off switch (SPST) to be used as an impact detector in the following configurations:

Figure 9.1a.   Using a microswitch as impact detector

Figure 9.1b.   Position sensor using one microswitch

Figure 9.1c.   Position sensor using two microswitches

Figure 9.1d.   Limit switch

Figure 9.1e.   Movement sensor

Figure 9.1f.   Acceleration/deceleration sensor

In the configuration of Fig. 9.1a, the switch can be placed on the front of a robot and used to sense when it bumps against any obstacle. Soft-touch types are ideal for this application, as it does not take much power to close them. In Figs. 9.1b–d,

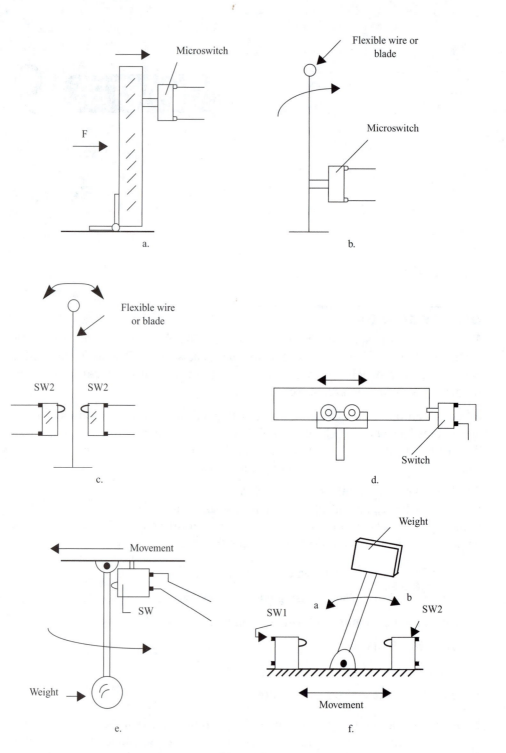

Figure 9.1

we show other applications in which switches can be used as sensors both for robotics and industrial automation. Figure 9.1f shows an interesting way that a sensor can be activated by braking or acceleration. This sensor can be used to reverse the direction of the drive motor in a robot or other vehicle when it suddenly strikes an obstacle.

Microswitches of the type used in industrial machines are also ideal for applications involving position sensors. They are small and sensitive, needing only a soft force applied for activation. Common switches can be converted into sensors using a variety of mechanisms. For example, a lever can increase the sensitivity of a microswitch (see Fig. 9.2.)

Pushbuttons can also be used as sensors, as shown in Fig. 9.3. It is important to choose a switch that can be easily closed by the mechanism that you are designing.

REED SWITCHES

A reed switch consists of a glass enclosure filled with an inert gas and blades with electrical contacts as shown in Fig. 9.4. Figure 9.4a shows a single-pole, single-throw (SPST) switch, and Fig. 9.4b shows a normally open/normally closed switch.

Figure 9.2

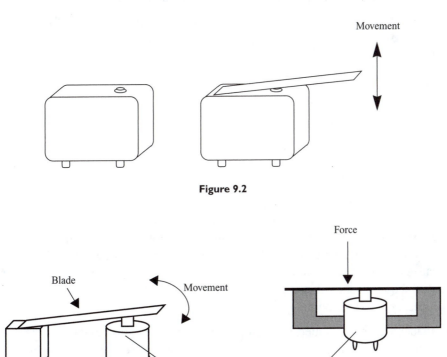

Figure 9.3

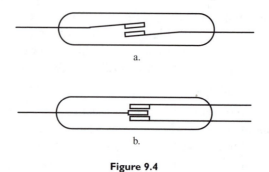

Figure 9.4

A magnetic field acts on the blades, causing them to touch each other, thereby closing the circuit. Reed switches are very sensitive but can't handle large currents. Typical currents are in the range of 50 to 500 mA. Since they can switch a circuit very quickly, they can be used in high-speed applications, sensing the rotation of a shaft or any mobile part of a mechanism.

By affixing a magnet to a wheel and placing a reed switch at an appropriate location, it is possible to produce a pulse with every revolution of the wheel. The pulses can be used to control the speed of the wheel or to excite a tachometer circuit.

Figure 9.5 shows reed switches used as sensors. Configurations are as follows:

Figure 9.5a.   Position sensor

Figure 9.5b.   Inclination sensor

Figure 9.5c.   RPM sensor

Figure 9.5d.   Limit switch

Figure 9.5e.   Proximity sensor

Figure 9.5f.   Reed relay

HOMEMADE SENSORS

Two or three blades arranged in close proximity, as shown in Fig. 9.6, form a mechanical sensor with many applications in robotics and mechatronics. The blades and the contacts can be salvaged from old relays or switches. The reader only needs to make sure that the contacts are in good working condition. Darkened or deformed contacts indicate burns that will degrade the efficiency of the sensor, so it is best not to use them.

The blades can be activated by many types of mechanisms. Figure 9.6 provides some suggested mechanisms that can act on this kind of sensor: (a) a simple blade SPST switch used as a position sensor, and (b) a three-blade SPDT switch used for the same purpose. A long spindle, for example, can be used as a position or obstacle sensor.

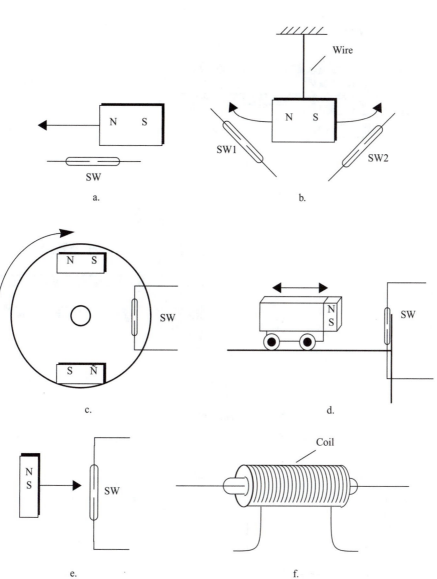

Figure 9.5

CONTACT CONDITIONERS

When a switch is closed, electrical contact is not established instantaneously. Mechanical contacts found in switches, relays, and sensors produce electromagnetic noise during their *setting time*. They bounce and make repeated contact during the first few milliseconds, as shown in Fig. 9.7.

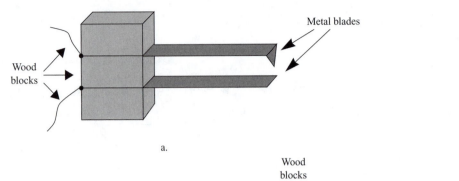

Metal blades

Wood
blocks

a.

Wood
blocks

NC

C

NO

Metal blades

b.

Figure 9.6

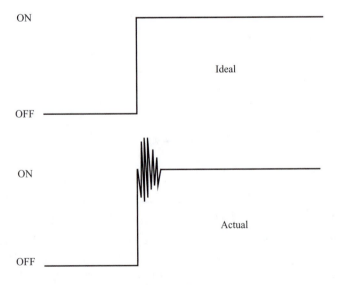

ON

OFF

Ideal

ON

OFF

Actual

Figure 9.7

Logic circuits (e.g., TTL and CMOS) and even some other electronic circuits are fast enough to recognize each and every bounce or noise pulse as a separate, intentional pulse or logic level change. You close the switch once and expect a single count, but the circuit will perceive the noise as a series of commands.

To avoid this problem when using conventional switches, reed switches, relays, or other on-off sensors with mechanical contacts, you need to install a signal conditioning circuit. Figure 9.8 shows the following "debouncing" configurations:

Figure 9.8a.   Simplest debouncing circuit, using a capacitor

Figure 9.8b.   Circuit using an RC network

Figure 9.8c.   Debouncing using TTL logic

Figure 9.8d.   Debouncing using CMOS logic

SWITCHES AS SENSORS

Many types of switches can be used as sensors or adapted to that task. Momentary-contact switches can be used as bump sensors, navigation feelers, and limit sensors. The most useful types are microswitches and pushbutton switches such as shown in Fig. 9.9.

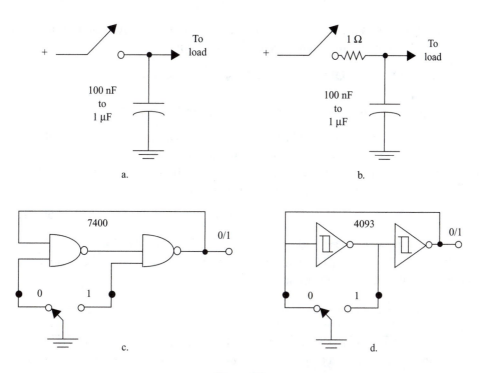

Figure 9.8

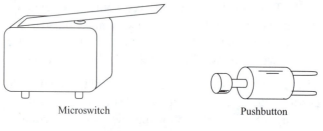

Microswitch   Pushbutton

Figure 9.9

HOMEMADE SENSORS

Many homemade sensors can be built by a typical robotics and mechatronics designer. Some examples are shown in Fig. 9.10, which shows the following types:

Figure 9.10a.   Incline sensor

Figure 9.10b.   Acceleration sensor

Figure 9.10c.   Bump sensor

Figure 9.10d.   Limit sensor

Figure 9.10e.   Two-position sensor

Since these sensors produce only short bursts of current when activated, appropriate signal conditioning circuits should be used with them.

PROGRAMMED OR SEQUENTIAL MECHANICAL SENSORS

Simple arrangements using wheels and screws can produce position sensors or programmed sensors that operate like the one shown in Fig. 9.11.

A cylinder covered with metal plating can be used as a position sensor. Each blade in the array touches the metal only at the points where no insulation is present. The designer can arrange the covered and uncovered positions of the cylinder so that the switches engage the proper signal at the proper time.

Other applications for sequential mechanical switches can be found in Chapter 15.

TTL AND CMOS SUITABLE FOR CONTACT CONDITIONERS

The following table lists TTL gates and inverters that can be used in contact conditioners and other logic applications.

Type	Function	Type	Function
7400	Quad two-input NAND gate	7414	Hex Schmitt trigger, inverting
7402	Quad two-input NOR gate	7486	Quad two-input exclusive/or gate
7404	Hex inverter		

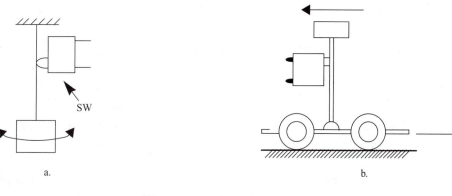

a.

b.

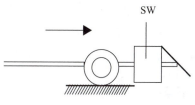

SW

c.

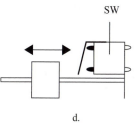

SW

d.

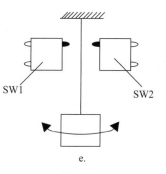

SW1

SW2

e.

Figure 9.10

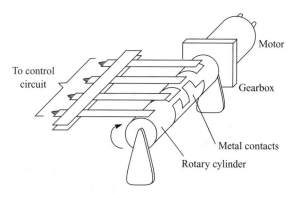

To control
circuit

Motor

Gearbox

Metal contacts

Rotary cylinder

Figure 9.11

Figure 9.12 shows the packages and pinouts for these ICs. Remember that any regular TTL output can sink 16 mA but deliver only 0.4 mA.

The following table lists CMOS gates that are suitable for our applications.

Type	Function	Type	Function
4001	Quad two-input NOR Gate	4010	Hex buffer (non inverting)
4011	Quad two-input NAND Gate	4093	Quad two-input NAND Schmitt trigger
4009	Hex buffer (inverting)		

Figure 9.13 shows the packages for these ICs. Remember that any CMOS output can sink or deliver currents, depending on the power supply voltage as given by the following table:

Power supply voltage	Output current (sink or deliver), typical
5 V	0.88 mA
10 V	2.25 mA
15 V	8.80 mA

RESISTIVE SENSORS

Resistive sensors can add sensory capabilities to any robotic, mechatronic, or artificial intelligence project. This section provides general information about

- Light sensors such as light-dependent resistors (LDRs) and cadmium sulfide (CdS) cells

- Negative temperature coefficient (NTC) resistors as temperature sensors

- Potentiometers used as a position sensors

- Conductive foam used as a pressure sensor, adding tactile sense to robots and mechanical arms

Touch sensors are also described in this section.

To sense the external world in robotic and mechatronic designs, we use *transducers*. A transducer is a device that converts one form of energy into another. For instance, a photocell used as sensor converts changes in light levels into electrical signals. Another type of transducer that acts on the ambient environment is the loudspeaker: it converts the electrical energy into sound (mechanical energy).

Robots and mechatronic devices can use many types of sensors. The following text describes some of them and also provides circuit blocks that can be used with these sensors.

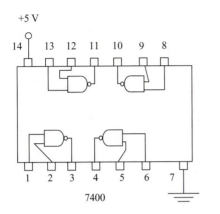

7400

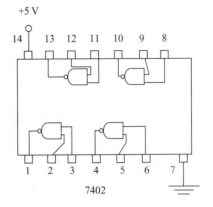

7402

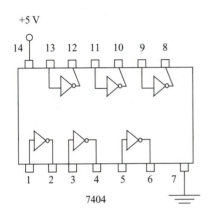

7404

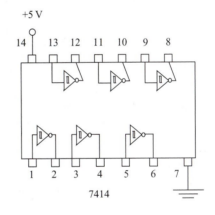

7414

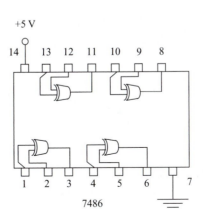

7486

Figure 9.12

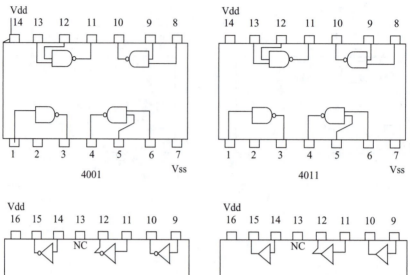

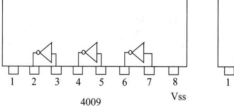

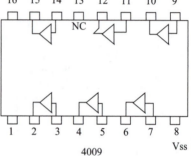

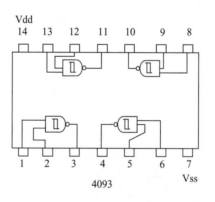

Figure 9.13

THE LDR OR CDS CELL

The light-dependent resistor (LDR) (also known as a CdS cell or photoresistor) is a component that changes its resistance in proportion to the amount of light that strikes its surface. In the dark, the LDR has a very high resistance—above 1 MΩ. This resistance will fall below 100 Ω in sunlight. Figure 9.14 shows the appearance and symbol used to represent an LDR.

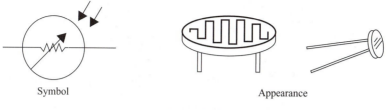

Symbol   Appearance

Figure 9.14

The LDR is a type of resistor, so the current can flow in both directions. Although the LDR is very sensitive, "seeing" light levels that our eyes cannot, it is a slow device. Quick light level changes cannot be detected by an LDR. The upper limit of the frequency response of an LDR is around 10 kHz. If you need to detect shorter light changes, you can use other sensors such as photodiodes and phototransistors.

Using an LDR is very easy, since it can directly bias semiconductor devices (transistors, SCRs, ICs, etc.). Figure 9.15 shows some circuits that employ an LDR as a sensor. All of these circuits, enumerated below, are of the ON/OFF type.

Figure 9.15a.   Using an NPN transistor

Figure 9.15b.   Using a PNP transistor

Figure 9.15c.   Using a power MOSFET

Figure 9.15d.   High gain using two transistors in Darlington configuration

Figure 9.15e.   High gain using complementary transistors

Figure 9.15f.   High gain using a Darlington power transistor

Figure 9.15g.   Very high gain using a comparator

Figure 9.15h.   Driving an SCR

The LDR devices can be used as "electronic eyes" in applications involving robotics, mechatronics, and artificial intelligence. As in human eyes, a lens can be added to change the performance of an LDR if required by the application. By placing a convergent lens in front of an LDR, we can pick up more light from one direction, increasing the sensitivity and adding directivity. Figure 9.16 shows how a lens can be used to add accomplish this.

THE NTC/PTC

Negative temperature coefficient (NTC) resistors, also called *thermistors* or *thermally sensitive resistors,* have a resistance that changes with temperature. The resistance of an NTC falls as the temperature increases.

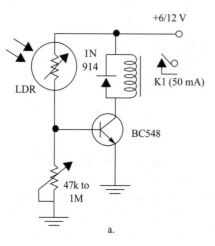

a.

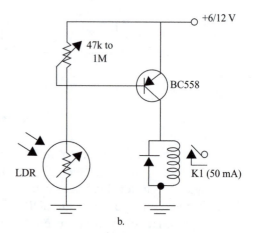

b.

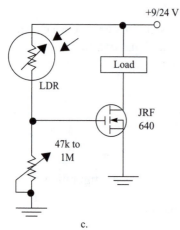

c.

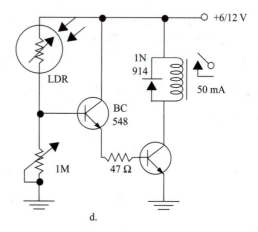

d.

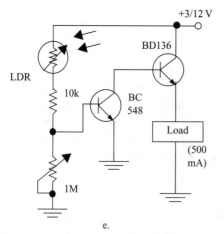

e.

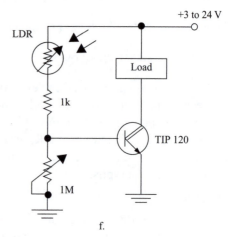

f.

Figure 9.15 *(continues)*

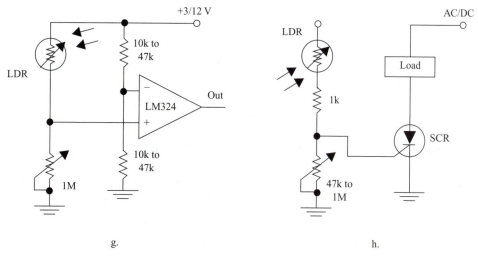

g.   h.

Figure 9.15 *(continued)*

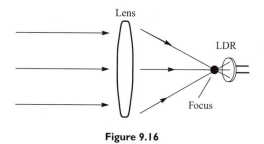

Figure 9.16

Another temperature-sensitive component is the positive temperature coefficient (PTC) resistor. In this component, resistance increases with temperature.

Figure 9.17 shows the symbol and appearance of common NTCs. These devices can be used to add temperature sensing to robots and other projects. They are specified by the resistance (in ohms) that they exhibit at a particular temperature (normally, ambient or 20°C). The values are typically in the range of several ohms to 100 kΩ.

Temperature sensors, like NTCs, are not particularly fast. A temperature sensor can't change its resistance quickly enough to keep up with brisk temperature changes. This is because the sensor's own temperature must change by absorbing or dissipating heat, and the reading will not be accurate until the sensor and ambient temperatures are equal. The reaction time is basically determined by the size and material of the NTC. Small NTCs work faster than larger ones and can sense quicker changes in the ambient temperature.

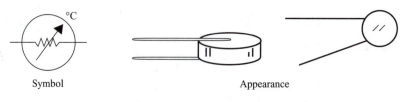

Symbol   Appearance

Figure 9.17

Since the electrical characteristics of NTCs (and PTCs) are the same as the those of an LDR, they can be used with the same circuits. We need only consider that they work in the opposite manner; i.e., if an NTC in a particular application turns a circuit on when the temperature rises, an LDR in the same circuit will turn it on when the temperature falls.

Considering the nominal resistance of NTC/PTC devices, they basically can be used in the same configurations adopted for LDRs as shown in Fig. 9.15. The only consideration is the need, in some cases, to change the bias in the resistors.

PRESSURE SENSORS

Simple pressure sensors can be fabricated using conductive foam such as the material used to protect ICs from electrostatic discharge. Figure 9.18 shows how an IC is protected by this foam.

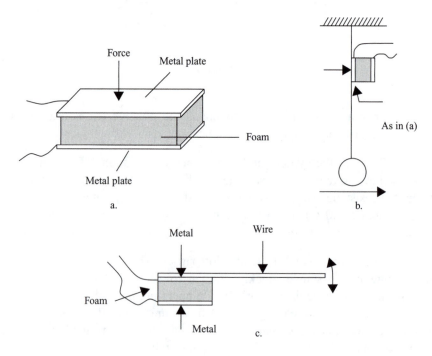

Figure 9.18

In this figure, we see some suggested sensor configurations as follows:

Figure 9.18a.   Pressure sensor

Figure 9.18b.   Position sensor

Figure 9.18c.   Acceleration/deceleration sensor

By placing the foam between two metal plates as shown in the figure, you create a pressure sensor. When pressure is exerted against the plates, the resistance of the foam changes (is reduced), and this reduction can be used to trigger a circuit. You can use this kind of sensor with the circuits recommended for LDRs and NTCs.

POTENTIOMETERS AS POSITION SENSORS

Common potentiometers can be adapted to function as position sensors. We need only connect some kind of mechanical device (e.g., a lever or wheel) to the shaft of a rotary potentiometer to convert the position of the device into an electrical signal. Figure 9.19 shows four ways to use a potentiometer as a sensor:

Figure 9.19a.   Position sensor

Figure 9.19b.   Two-coordinate position sensor

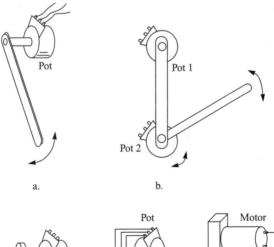

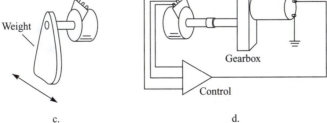

Figure 9.19

Figure 9.19c.   Incline sensor

Figure 9.19d.   Servo feedback sensor

Potentiometers with values in the range of 10 kΩ to 1 MΩ are commonly used. Any circuit that is suitable for use with resistive sensors can be used with potentiometers. These circuits will deliver a voltage that is proportional to the angle of the sensor (or the position of the cursor, if a slide type is used).

TOUCH SENSORS

A touch sensor can be formed by two metal plates that are in close proximity but not touching. If you place your finger on the two plates so that you are touching both at the same time, as shown in Fig. 9.20, an electric current can flow across your skin as an electric signal. Since the resistance of your skin is very high (100,000 Ω or more), the current flowing through this sensor will not be sufficient to drive any common circuits. Amplification stages are needed.

It is extremely important to keep the current low if high voltage is used in this sensor. If the current is too high, you can experience a shock. Therefore, the main considerations in designing a touch sensor of this type are that it (a) must be isolated from the AC power line and (b) must employ current-limiting resistors.

Another example of a touch sensor is shown in Fig. 9.21. If one plate is removed and connected to ground, you merely have to touch the other plate to trigger the circuit. The current will flow through the body, activating the circuit.

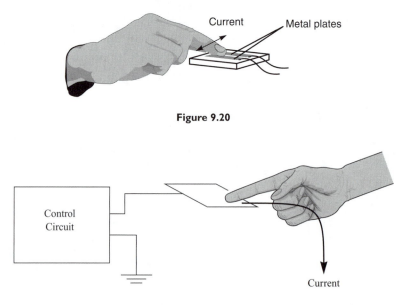

Figure 9.20

Figure 9.21

Figure 9.22 shows the following circuits that use touch sensors:

Figure 9.22a.   Using bipolar transistors—this circuit can drive relays and loads as described in Chapter 5.

Figure 9.22b.   Using power MOSFETs—this circuit can also drive relays as described in Chapter 5.

Figure 9.22c.   Using CMOS ICs.

Figure 9.22d.   Using operational amplifier or comparators—the gain is fixed by R1.

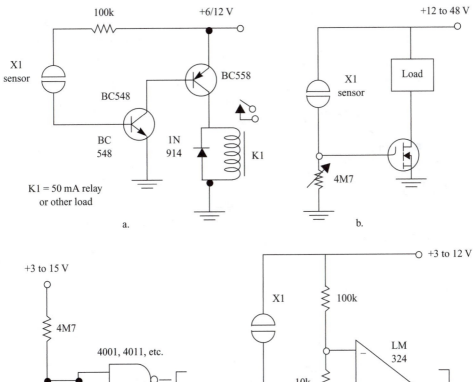
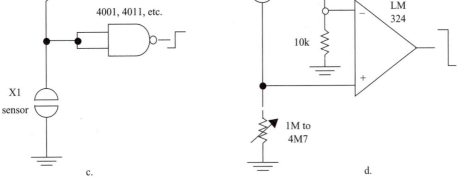

Figure 9.22

PIEZOELECTRIC SENSORS

Piezoelectric ceramic devices such as found in buzzers and microphones can be used as sensors in many applications. They can detect sounds, bumps, cracks, and deformations. Figure 9.23 shows some circuits using piezoelectric sensors. The sensors are based on the type of small buzzers that are primarily intended for signaling applications. Configurations are as follows:

Figure 9.23a.   Using transistors

Figure 9.23b.   Using an operational amplifier/comparator

Figure 9.23c.   Using an audio amplifier

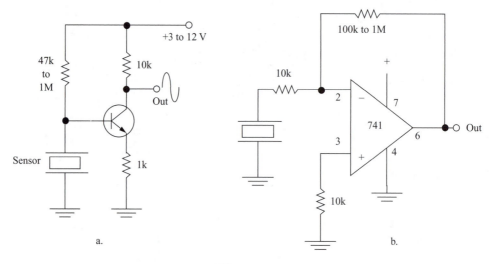

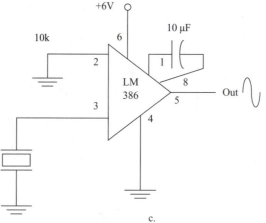

Figure 9.23

HUMIDITY

The presence of water or humidity can be detected using the same circuits as suggested for operation with NTCs, PTCs, and LDRs. The only difference is that the bias resistor must be changed in some cases to adapt the resistive characteristics of the sensor to obtain the desired sensitivity. Figure 9.24 shows the following configurations of humidity/water sensors:

Figure 9.24a.   Two metal screens and a piece of tissue with some salt

Figure 9.24b.   Two wires with bare ends

Figure 9.24c.   Semiconductor-type humidity sensor

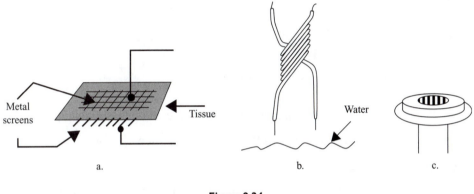

Figure 9.24

HALL EFFECT SENSORS

Hall effect sensors detect magnetic fields. They are classified as resistive sensors, and their resistance depends on the action of an external magnetic field. They are very fast and sensitive. Figure 9.25 shows three examples of this type of sensor.

Figure 9.25a.   Movement sensor

Figure 9.25a.   Speed sensor style 1

Figure 9.25a.   Speed sensor style 2

SOUND

Microphones can be used to detect sounds. Microphone signals need amplification, since their output signals are not strong enough to drive a relay or other load. Figure 9.26 shows the following circuits using microphones as sound sensors:

Figure 9.26a.   Low-impedance circuit using a transistor

Figure 9.26b.   High-impedance circuit using a bipolar transistor

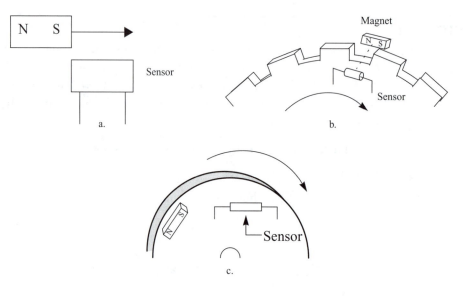

Figure 9.25

Figure 9.26c.   Medium/high impedance circuit using an operational amplifier

Figure 9.26d.   Circuit using the LM386 (National Semiconductor)

HOW TO USE RESISTIVE SENSORS

When using resistive sensors, the designer must keep in mind two important questions.

- How much does the resistance change under the intended operating conditions?

  The resistance range of a sensor is important in the design or choice of the circuit to which it is connected. If a sensor's resistance changes by only a few ohms when excited, the circuit must have special characteristics to react to this small change.

  LDRs change their resistance over a wide range of values, so they are easy to use. NTCs, depending on the type, may also provide a wide resistance range, but other types (e.g., pressure sensors) do not.

- Is the sensor's output current sufficient to drive a circuit without amplification?

  If the sensor experiences a large resistance drop when excited, the current may be sufficient to drive the attached circuit. In general, to drive a block with a bipolar transistor, the resistance in the low-value condition must be below 50,000 Ω. If a Darlington transistor, CMOS ICs, or two-stage circuits are used, this resistance can be higher—in the range of 200 kΩ to 1 MΩ. As a general

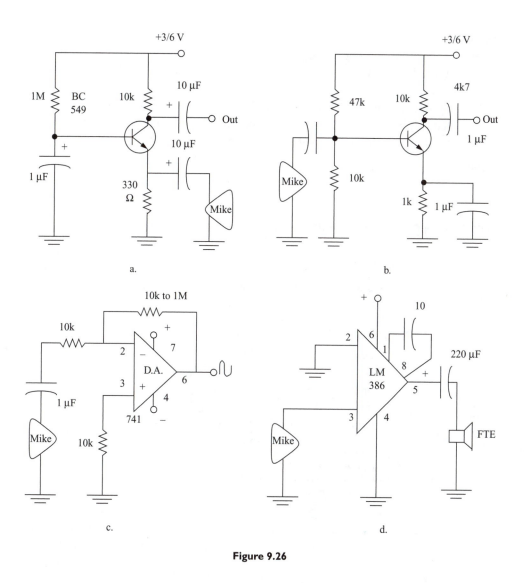

Figure 9.26

rule, if your sensor does not trigger the circuit, try a block that offers higher sensitivity.

SENSOR CIRCUITS—MONOSTABLE/ASTABLE

The circuits described in this section are suitable for sensor applications.

Monostable circuits can provide a constant-duration signal when triggered by on/ off sensors. *Astable* circuits can act as analog-to-digital (ADC) converters, delivering signals with a frequency that is proportional to the value of the analog quantity detected by the resistive sensor.

MONOSTABLE 555

When connected as a monostable multivibrator, the 555 needs an external trigger command applied to the trigger input (pin 2) to start the action. This is normally done by leaving the trigger input in a positive state and by momentarily setting it to ground. By this action, the output goes to the high logic level (positive) for a time interval that can be calculated by the following formula. Figure 9.27 shows the 555 in the monostable configuration. (See also Table 9.1.)

Formula:

$$T = 1.1 \times R \times C \tag{9.1}$$

where T = period or time on in seconds (s)
R = resistance in ohms (Ω)
C = capacitance in farads (F)

Table 9.1   Limit Values for the Monostable 555

Parameter/component[1]	Limit value
R max	3 MΩ
R min	1 k Ω
C max	2 000 μF[2]
C min	500 pF
T_r (max)	1/4 T
I_{out} (drain or source)	200 mA
Vcc	18 V

[1] I_{out} = maximum output current, Vcc = power supply voltage, and T_r = trigger pulse duration in seconds.
[2] Depends on leakage.

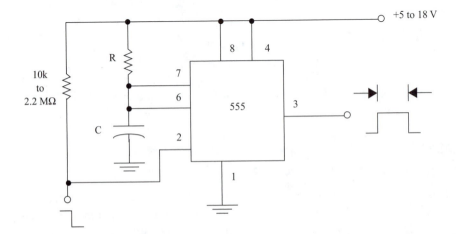

Figure 9.27

ASTABLE 555

The 555 integrated circuit can be used in two basic configurations: as an astable multivibrator producing squarewave signals in a frequency range up to 500 kHz, and as a monostable multivibrator or timer.

Figure 9.28 shows the basic astable configuration. The following formulas are used for calculations involving the 555 astable.

Formulas:

Charge time of capacitor C (output high):

$$T_h = 0.693 \times C \times R_1 + R_2 \tag{9.2}$$

where T_h = time the output is high in seconds (s)
R_1 and R_2 = resistances in ohms (Ω)
C = capacitance in farads (C)

Discharge time of capacitor C (output low):

$$T_L = 0.693 \times R_2 \times C \tag{9.3}$$

where T_L = time the output is low in seconds (s)
R_2 = resistance in ohms (Ω)
C = capacitance in farads (F)

Period:

$$T = 0.693 \times (R_1 + 2 \times R_2) \times C \tag{9.4}$$

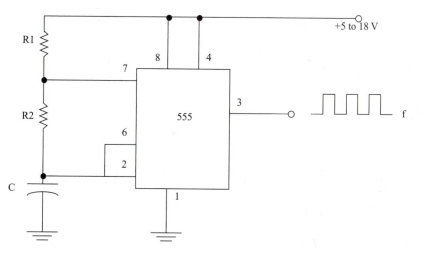

Figure 9.28

where T = period in seconds (s)
          R_1 and R_2 = resistances in ohms (Ω)
                  C = capacitance in farads (C)

Frequency:

$$f = \frac{1.44}{(R_1 + 2 \times R_2) \times C}$$ (9.5)

where f = frequency in hertz (Hz)
          R_1 and R_2 = resistances in ohms (Ω)
                  C = capacitance in farads

Duty cycle:

$$Dc = \frac{T_h}{T_L}$$

$$Dc = \frac{R_1 + R_2}{R_2}$$ (9.6)

where Dc = duty cycle (0 to 1)
              T_h = time the output is high in seconds (s)
              T_L = time the output is low in seconds (s)
     R_1 and R_2 = resistances in ohms (Ω)

Obs.: Multiply by 100 if you want the result to be a percentage value.

Table 9.2   Recommended Limit Values for the Astable 555

Component	Limit value
R1 + R2	max. 3 MΩ
R1	min. 1 kΩ
R2	min. 1 kΩ
C	min. 500 pF
C	max. 2,200 μF[1]
F	max. 1 MHz
I_{out} (drain or source)	200 mA
Vcc	18 V

[1] Depends on the leakage.

Obs.: There is a CMOS version of the bipolar 555 called the TLC7555, with upgraded characteristics including higher output current and upper frequency limit.

CHAPTER 10

Power Electronics

BIPOLAR POWER TRANSISTORS

Bipolar transistors are the most important of all semiconductor devices. They are used in nearly all electronic equipment as well as in mechatronic projects. They can be used as switches or amplifiers. Transistors with high-current capabilities are very important elements in mechatronic designs, as we explain in this chapter.

TRANSISTOR STRUCTURES

Bipolar transistor can be found with two basic structures as shown in Fig. 10.1. They are formed by three pieces of semiconductor material (generally silicon) forming two junctions in the same crystal. Although the structure can be compared

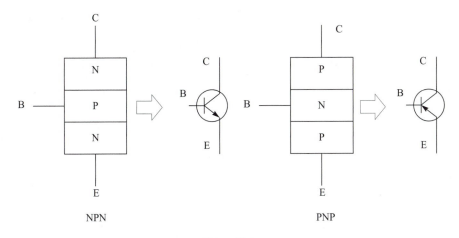

Figure 10.1

to two diodes placed back to back, the fact that they exist in the same crystal means that the events occurring at one junction can influence events at the other.

The possibility of using the transistor for control functions makes it an important element in mechatronic applications. From the two types of semiconductor structures, we obtain two types of transistors, NPN and PNP.

Observe in the figure that terminals or leads are attached to each piece of semiconductor material, corresponding to three regions: base (B), emitter (E), and collector (C).

HOW THEY WORK

As shown in Fig. 10.2, a current flowing between the emitter and collector can be controlled by a current flowing into the base. The direction of the current depends on the type of the transistor. Currents in an NPN transistor flow in the opposite direction of those flowing in PNP transistors, as shown in Fig. 10.3.

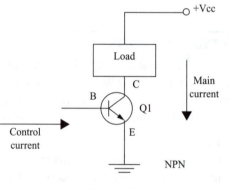

Figure 10.2

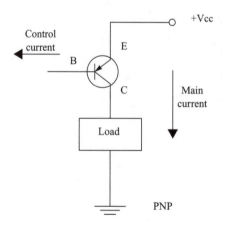

Figure 10.3

The most important concept to understand about transistor function is that a small current flowing between the base and the emitter can control a large current between the collector and emitter. This means that the transistor has a "gain" or "amplification factor" (also referred to as *beta*, or β). If a current of 1 mA applied to the base causes a 100 mA current to flow through the collector, the transistor has a beta of 100.

A transistor can be used in two different ways, as illustrated by the "characteristic curve" shown in Fig. 10.4. When operating in the linear region, small changes in the base current will cause large changes in the collector current. In this type of operation, the transistor acts as a signal amplifier. If the base current is an audio signal, the input signal will be amplified in the collector.

When operating in the saturation region, the transistor acts as a switch. A current in the base "turns on" the transistor, allowing a large current flow to the collector as shown in Fig. 10.5.

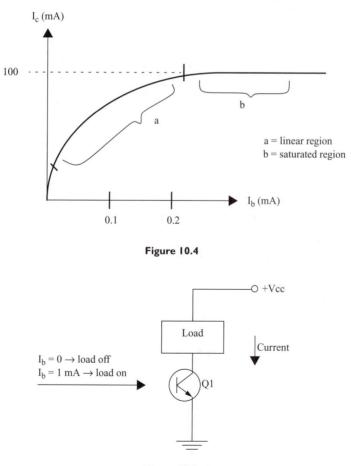

Figure 10.4

Figure 10.5

The use of the transistor as a switch is the most important application in mechatronics, because this element can be used to control large loads such as motors, solenoids, shape memory alloys, relays, and other devices using small currents. See Chapters 4 through 6 for application examples.

SPECIFICATIONS

For various applications, it is possible to find transistors of different sizes and formats with gains ranging from 5 to 1000. Transistors are also classified according to the amount of current they can handle.

Small transistors, signal transistors, and general-purpose transistors can control only small currents. They are intended for applications in which a low-power signal may be amplified or controlled. Typical units cannot control currents above a few hundred milliamperes.

An important type of transistor, known as the *power transistor,* can control large currents such as the ones drawn by motors, solenoids, SMAs, and other loads that are commonly found in mechatronic designs. Figure 10.6 shows some common power transistors.

In operation, *bipolar power transistors* generate heat, so they need to be mounted on heatsinks. The size and the type of heatsinks depends on the amount of power generated by the transistor in a particular application.

When using a transistor, we must consider certain electrical, mechanical, and thermal characteristics as described below.

Maximum Voltages

A transistor has certain voltage limits that cannot be exceeded. These are as follows.

$V_{ce(max)}$ The maximum voltage that can be applied between collector and emitter.

$V_{ceo(max)}$ Same as above but with the base open.

$V_{cb(max)}$ The maximum voltage that can be applied between collector and base.

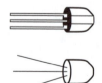

Low power

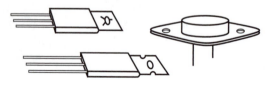

Medium and high power

Figure 10.6

$V_{cbo(max)}$ Same as above but with the emitter open.

$V_{eb(max)}$ The maximum voltage that can be applied between emitter and base.

$V_{ebo(max)}$ Same as above but with the collector open.

Maximum Current

The maximum current that can flow through the collector terminal is referred to as $I_{C(max)}$. Other current ratings, such as base current, I_b, and emitter current, I_e, also may be specified.

Gain

As mentioned previously, a small current flow into the base causes a large current flow between the emitter and collector. The factor to which the collector current is larger than the base current is the transistor gain or beta (β) factor. The gain may also be indicated by a hybrid parameter called *hFE* or h_{FE}.*

Maximum Heat

When conducting a current, a transistor generates heat that must be dissipated into the air. The maximum amount of heat produced by a transistor is given by its maximum dissipation power, $P_{d(max)}$.

Transition Frequency

The gain of a transistor drops when it is working with high-speed signals. There is a frequency at which the gain falls to unity. It is called the *transition frequency* or f_T. At this frequency, a variation of 1 mA in base current causes a variation of 1 mA in the collector current, and the transistor can no longer operate as an amplifier.

Formula:

The power generated by a transistor can be calculated by

$$P_d = V_{ce} \times I \qquad (10.1)$$

where P_d = power produced (and dissipated) by the transistor (W)

         V_{ce} = voltage between the emitter and collector when the transistor is conducting (V)

         I = current flowing through the transistor (A)

Obs.: Common bipolar silicon transistors have a typical V_{ce} when conducting (saturated) of between 1.2 and 2.5 V.

---

*This has been defined as the "static value of the forward current transfer ratio in common-emitter configuration."

Formula:

Transistor gain can be computed as

$$h_{FE} = I_c/I_b \tag{10.2}$$

where h_{FE} = gain (β)

I_c = collector current (mA)

I_b = base current that causes the collector current (mA)

Transistor Identification Codes

The three standard transistor marking schemes are described below.

1. Joint Electron Device Engineering Council (JEDEC) codes. These take the form

   digit, letter, serial number, [suffix]

   The letter is always N, and the first digit is one less than the number of leads. This is a 2 for transistors except for field-effect transistors (FETs), which use the 3N, and except for 4N and 5N, which are reserved for optocouplers. The serial number runs from 100 to 9999 and tells us nothing about the transistor except its approximate time of introduction.

   The (optional) suffix indicates the gain (h_{FE}) group of the device:

   A = low gain

   B = medium gain

   C = high gain

   No suffix = ungrouped (any gain)

   See the data sheet for the actual gain spread and groupings. The reason for gain grouping is that low-gain devices are cheaper than the high-gain devices, resulting in savings for high-volume users. Examples are 2N3821, 2N2222A, and 2N2904.

2. Japanese Industrial Standard (JIS) codes. These take the form

   digit, two letters, serial number, [suffix]

   Again, the digit is one less than the number of leads.

   The letters indicate the application area and type of device according to the following codes:

SA: PNP HF transistor	SK: N-channel FET/MOSFET
SB: PNP AF transistor	SM: triac
SC: NPN HF transistor	SQ: LED

SD: NPN AF transistor	SR: rectifier
SE: diodes	SS: signal diodes
SF: thyristors	ST: avalanche diodes
SG: Gunn devices	SV: varicaps
SH: UJT	SZ: zener diodes
SJ: P-channel FET/MOSFET	

The serial number runs from 10-9999. The (optional) suffix indicates that the type is approved for use by various Japanese organizations.

Note: Since the code for transistors always begins with 2S, it is sometimes omitted so, for example, a 2SC733 would be marked C733. Examples include 2SA1187, 2SB175, and 2SC733.

3. Pro Electron codes.[*]

These take the form,

two letters, [letter], serial number, [suffix]

The first letter indicates the material according to the following:

A = Ge

B = Si

C = GaAs

R = compound materials

The second letter indicates the device application as indicated below:

A: diode, RF	N: optocoupler
B: variac	P: radiation-sensitive device
C: transistor, AF, small signal	Q: radiation-producing device
D: transistor, AF, power	R: thyristor, low power
E: tunnel diode	T: thyristor, power
F: transistor, HF, small signal	U: transistor, power, switching
K: Hall effect device	Y: rectifier
L: transistor, HF, power	Z: zener, or voltage regulator diode

The third letter indicates whether the device is intended for industrial or professional rather than commercial applications. It is usually a W, X, Y, or Z. The serial number runs from 100-9999. The suffix indicates the gain grouping, as for JEDEC. Examples are BC548A, BAW68, BF494, and BFY51.

---

*Pro Electron is a nonprofit organization operating as an agency of the European Electronic Component Manufacturers Association (EECA), Brussels, Belgium.

Beyond JEDEC, JIS, and Pro Electron, many manufacturers often introduce their own designations for commercial or other reasons. Examples of specific prefixes are:

MJ: Motorola power, metal case

MJE: Motorola power, plastic case

MPS: Motorola low power, plastic case

RCA: RCA

RCS: RCS

TIP: Texas Instruments power transistor, plastic case

TIPL: Texas Instruments planar power transistor

ZT: Ferranti

ZTX: Ferranti

Examples include ZTX302, TIP31A, MJE3055, TIP31, and TIP120.

HOW TO USE BIPOLAR POWER TRANSISTORS

The way the transistor are biased and how the signals to be amplified or controlled are applied to and drained from a circuit determine the three basic connection configurations shown in Fig. 10.7. Table 10.1 shows come common characteristics of these circuits.

Table 10.1 General Characteristics of Transistor Circuits

Common emitter	Common base	Common collector
Large voltage gain	Large voltage gain	Near unity voltage gain
Large current gain	Near unity current gain	Large current gain
Highest power gain	Medium power gain	Lowest power gain
Low input resistance	Very low input resistance	High input resistance
High output resistance	Very high output resistance	Low output resistance

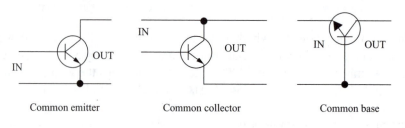

Common emitter   Common collector   Common base

Figure 10.7

Figure 10.8 shows how a transistor is used in the two basic operational modes (linear and saturated). In Fig. 10.8a, the load is placed between the power supply and the collector of an NPN transistor. The resistor in the base determines the amount of base current. This resistor is called the *biasing resistor*. When we close S1, the base current causes the current to flow through the load.

If, as shown in Fig. 10.8b, a variable current (such as corresponding to an audio signal) is applied to the base, small variations in this current will produce large variations in the collector current.

The way the transistor is biased, and how the signals to be amplified are applied to and drawn from a circuit, determine the three basic transistor configurations. These basic arrangements are shown in Fig. 10.9.

In Fig. 10.9a, we have the common emitter configuration, producing higher voltage and current gain (power gain). In Fig. 10.9b, we have the common emitter configuration giving a high current gain but low voltage gain, and in Fig. 10.9c, we have the common base configuration. Depending on the application, one configuration will offer better performance and will be preferred.

Normally, more than one transistor is used in a particular application. Each transistor is the center of a stage, and the amplified signals must pass from one stage to other, being amplified at each step until we reach the desired level.

The basic function of power transistors in mechatronics is to control high current loads, as shown in Fig. 10.10. Here, the value of resistor R is chosen so as to deliver enough current to the base to produce the current needed by the load, with consideration given to the resistor gain.

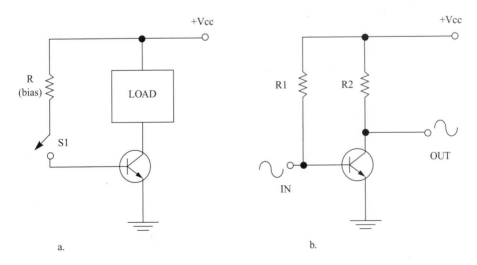

Figure 10.8

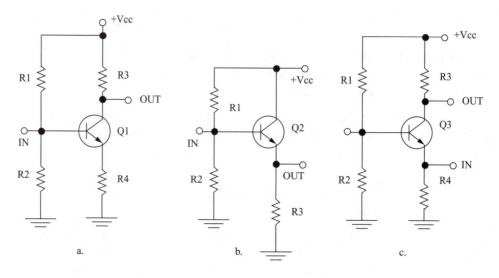

Figure 10.9

The configurations are as follows:

Figure 10.10a. Common NPN

Figure 10.10b. Common PNP

Figure 10.10c. Complementary I

Figure 10.10d. Complementary II

Figure 10.10e. Darlington NPN

Figure 10.10f. Darlington PNP

Formula:
An approximate formula to determine the value of resistor R is

$$R = \frac{Vcc - 0.6}{Ic} \times \beta \tag{10.3}$$

where R = value of the resistor (Ω)
       Vcc = power supply voltage (V)
       I_c = collector or load current (A)
       β = gain of the transistor

Obs.: (a) The value calculated is the maximum needed to saturate the transistor. For practical purposes, use a value 20 to 30 percent lower than calculated. (b) Use the minimum β (gain) of the transistor under consideration.

If the transistor is driven from logic sources (TTL or CMOS), see Chapter 16.

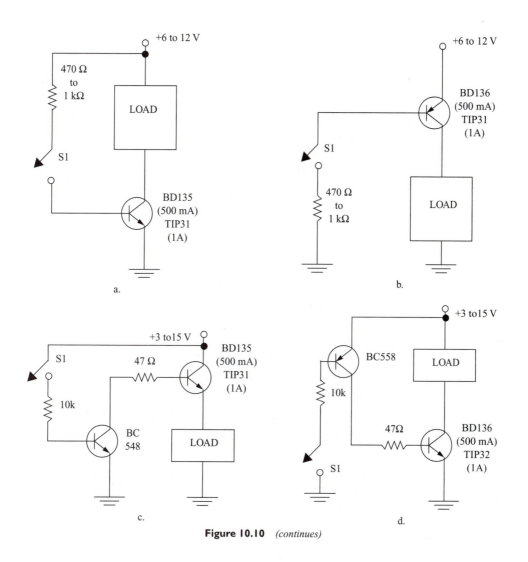

Figure 10.10 *(continues)*

TEST METHODS

The primary test that we want to make is a measurement of the voltages at the transistor leads. For that, we use a multimeter. As a general rule, the collector voltage will be higher than the emitter voltage in an NPN transistor, and vice versa in a PNP transistor (see Fig. 10.11).

The base voltage is 0.6 to 0.7 V above the emitter voltage in an NPN silicon transistor, and the same amount below the emitter voltage in a PNP silicon transistor. In old germanium transistors, the difference is 0.2 V. If the voltages are otherwise, the biasing resistor or other elements surrounding the transistors can develop problems.

A *static test* of a transistor (i.e., a test when the device is not installed in a circuit) can be made by measuring the resistance between the leads (E–B–C) using a multi-

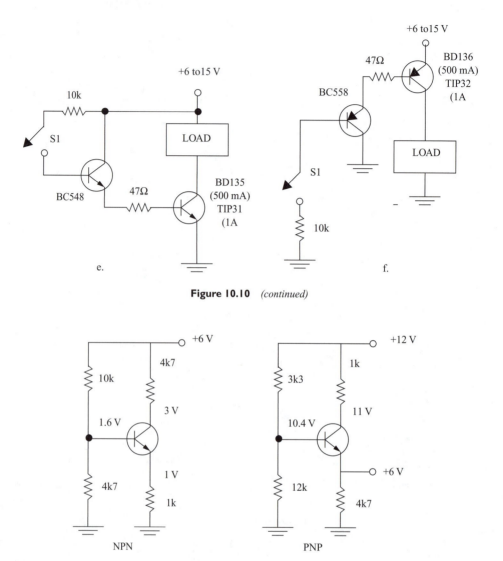

Figure 10.10   *(continued)*

Figure 10.11

meter. It is called a *junction test* when we verify that the two junctions (base-emitter and base-collector) are functional.

The basic test consists of observing whether the two "virtual" diodes that correspond to the two junctions are good. Therefore, you just need to adjust the multimeter to a low-resistance scale and make six measurements to determine if the transistor is functioning. Figure 10.12 shows how to perform these measurements.

When placing the probes between base/collector and base/emitter, one measurement must be of a high resistance and the other of a low resistance. Both measure-

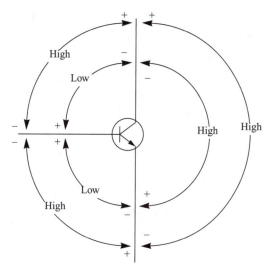

Figure 10.12

ments, when you have placed the probes between the collector and emitter, must result in a high resistance reading. Anything else indicates that the transistor has problems (a short or open).

If we expect high resistance (i.e., more than 1 MΩ) but obtain a value between 20,000 and 500,000 Ω, we say the transistor *presents losses*. It also qualifies as a bad transistor.

BIPOLAR POWER DARLINGTONS

Two transistors wired as shown in Fig. 10.13 form a Darlington stage or a Darlington amplifier. This stage acts as a *super transistor,* meaning that the Darling stage gives us a gain corresponding to the product of the gains of the individual transistors. For instance, if two transistors with gains of 100 ($\beta = 100$) are placed together

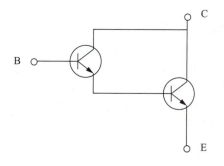

Figure 10.13

to form a Darlington pair, the final gain will be 100×100 or 10,000! They will function as a single transistor with a gain of 10,000.

Darlington transistors can be manufactured by placing two transistors on the same silicon chip and packing them inside a common enclosure to form the super transistor. Darlingtons are useful where high gain is required, such as in solenoid, motor, or relay drivers; lamp drivers; high-power audio amplifiers; and so on.

The symbols used to indicate Darlington transistors are shown in Fig. 10.14. As with common bipolar transistors, they are available in NPN and PNP versions.

Just by appearance, it is not possible to tell if a component is a common or a Darlington transistor, as these and other semiconductors use the same package. Identification is accomplished by part number, which is usually printed on the package.

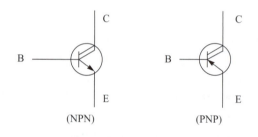

Figure 10.14

SPECIFICATIONS

A common series of Darlington transistors found in many electronic devices, and even in electrical appliances, is the TIP series of Darlington transistors. These were first manufactured by Texas Instruments but are now offered by many other manufacturers. This series includes various NPN and PNP Darlingtons that are designed to control currents from 1 to more than 10 A and voltages up to 100 V and beyond.

Some Darlington NPN power transistors in this series are the TIP110, TIP111, and TIP112, for 2-A collector currents. The corresponding PNP types are the TIP115, TIP116, and TIP117.

Important: Not all transistors whose codes begin with "TIP" are Darlingtons!

Manufacturers' data sheets normally offer much information about the transistor, including many graphs and numerical data about their performance. For the electrician, or even electronic technicians who only need information about how to use or replace a Darlington transistor, the important specifications are the same as for common bipolar transistors:

    a. Collector-emitter voltage (V_{ce})

    b. Collector current (I_c)

c. Gain (hFE)

d. Dissipation Power (P$_d$)

e. Transition frequency (f$_T$)

HOW TO USE DARLINGTON POWER TRANSISTORS
Darlington transistors are used to drive high current loads as solenoids, electromagnets, motors, lamps, as shown by the example circuit in Fig. 10.15. Figure 10.15a uses an NPN Darlington, and Fig. 10.15b employs a PNP version.

The value of resistor R can be calculated using Eq. (10.3). For practical purposes, with the voltages and currents as shown in the examples, the resistor will be in the range of 2.2 to 10 kΩ.

TEST METHODS
Darlington transistors can be tested using the techniques described for common bipolar transistors. Using a multimeter, you can measure the resistances between electrodes (base, emitter, and collector) according to the type (NPN and {PNP). See *Test Methods,* p. 205, for details. Keep in mind that some multimeters have a built-in transistor test function that can be used with Darlingtons. Table 10.2 lists some common Darlington transistors and their characteristics. Corresponding packages are illustrated in Fig. 10.16.

POWER MOSFETS
The metal-oxide semiconductor field-effect transistor (MOSFET) is derived from the common FET but with some variations in the basic structure and electrical characteristics.

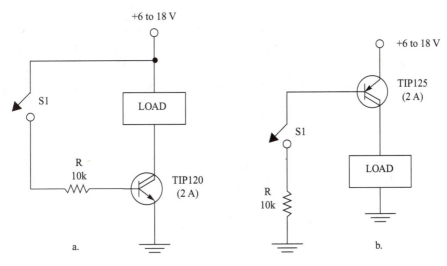

Figure 10.15

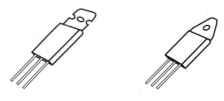

Figure 10.16

Table 10.2   Common Darlington Transistors (TIP Series)

NPN	PNP	I_c (A)	V_{ceo} (V)	h_{Fe} (min)	Package (see Fig. 10.16)
TIP110	TIP115	2	60	1000	(a)
TIP111	TIP116	2	80	1000	(a)
TIP112	TIP117	2	100	1000	(a)
TIP120	TIP125	5	60	1000	(a)
TIP121	TIP126	5	80	1000	(a)
TIP122	TIP127	5	100	1000	(a)
TIP140	TIP145	10	60	1000	(b)
TIP141	TIP146	10	80	1000	(b)
TIP142	TIP147	10	100	1000	(b)
TIP640	TIP645	10	60	1000	(c)
TIP641	TIP646	10	80	1000	(c)
TIP642	TIP647	10	100	1000	(c)

As shown by the structure in Fig. 10.17, the MOSFET uses a thin layer of metal oxide (silicon oxide) to isolate the substrate from the gate region, rather than a junction as found in the junction field-effect transistor (JFET). However, although there are structural differences, the operation of a MOSFET is the same: a voltage

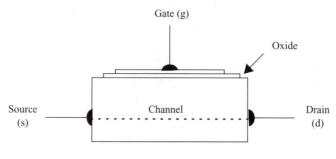

Figure 10.17

applied at the gate can control the amount of current across the device or the current between drain and source.

Warning: The layer that isolates the gate region from the channel is very thin, making the device extremely sensitive to static discharge. If you have built up a static charge in your body, and you touch any terminal of the device, a spark between the gate and channel can be produced that will destroy the isolating layer and therefore the device.

Figure 10.18 shows the schematic symbols used to represent the two types of MOS-FETs commonly found in electronic circuits. The MOSFET can also be found in a dual-gate version. The amount of current across the device can be controlled independently by the voltages applied to each of the two gates. This device is particularly well suited for applications in which two signals must be mixed.

Common MOSFETs are low-power devices, and their overall appearance is the same as bipolar transistors. To identify a particular device, you need a schematic diagram, a data sheet with the pertinent information, or the part number (generally printed on the device).

Power FETs are MOSFET versions that can be used to control large currents. They are very common today in digital and analog applications, as they offer advantages over bipolar transistors in many cases. Power FETs (also called power-MOSFETs) can control currents as high as 50 A with voltages exceeding 500 V.

The structure and the operating principle of a power MOSFET is the same as that of the common MOSFET. The differences lie in the size and shape of the silicon chip used in manufacturing the device.

Several technologies are used to produced these devices, from which we derive the terms V-FETs, power-FETs, D-FETs, TMOS, and so on.

A power FET operates in the same manner as a common MOSFET. A voltage applied to the gate can control the amount of current that flows across the device. High-current devices (e.g., solenoids, motors, electromagnets, relays, heaters, lamps, etc.) can be controlled by power FETs in the same kind of circuit stages that we would use for power bipolar transistors and Darlington transistors.

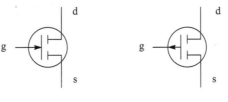

N-type   P-type

Figure 10.18

The main advantage to using power-MOSFETs is the low resistance that they display when conducing a current (named R_{ds} for drain source resistance). As the amount of heat produced by the device depends on this resistance value, power-FETs can control much larger currents than equivalent bipolar transistors while generating less heat.

Figure 10.19 shows that power FETs can be found in two types—N- and P-channel. They look the same as power bipolar transistor, as both devices use the same packages.

N-channel MOSFETs conduct the main current when a positive voltage is applied to the gate, and P-channel MOSFETs conduct when a negative voltage is applied to the gate. As with any other power transistor, power MOSFETs must be mounted on heatsinks to dissipate the heat generated while in operation.

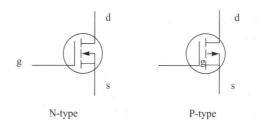

N-type   P-type

Figure 10.19

SPECIFICATIONS

Power MOSFETs can be identified by part number. The popular IRF series is produced by many manufacturers, including International Rectifier, Fairchild, Motorola, and others. All devices in this series have names beginning with IRF, e.g., IRF640, IR720, IRF630, IRF730, and so on.

Using the device's part number, it is possible to find the electrical characteristics of a power MOSFET in handbooks, on the Internet, or on product data sheets. The principal characteristics are described below.

Maximum Voltage between Drain and Source

This refers to the maximum voltage that can be applied to the device without burning it out. Common types have voltages ranging from 50 to 1,000 V. This specification may be indicated as $V_{ds(max)}$.

Maximum Drain Current

This is the highest current that can be controlled by the device. It is abbreviated Id and can range from 1 to 50 A. This specification is also abbreviated I_d or $I_{d(max)}$.

Dissipation Power

This indicates the amount of heat that the device can dissipate into the environment. It is measured in watts and can range from 10 to 250 W.

Resistance between Drain and Source in the On State

This is one of the most important power FET specifications. The amount of operational heat generated by the device depends on this resistance. Multiplying this resistance times the square of the current gives the generated power. It is easy to see that the lower this resistance, the more efficient the device will be when controlling high currents. This resistance should be less than 1 Ω in most common devices. It is commonly indicated by rDS or R_{ds} and is given in units of ohms (Ω).

HOW TO USE POWER MOSFETS

Power MOSFETs can replace common bipolar transistors in the control of high current loads. Figure 10.20 shows some circuits that employ power MOSFETs.

Figure 10.20a.   On-off switch for low-current switches

Figure 10.20b.   Control stage

Figure 10.20c.   Timer

Figure 10.20d.   Circuit driven by CMOS logic

Since power MOSFET are very high input impedance devices, they need almost no power to be driven. The voltage to turn on a MOSFET is typically between 1 and 2 V.

TEST METHODS

The simplest test is accomplished by measuring the resistances between the drain and source and between the gate and drain. Either one may be very high.

Obs.: Some power MOSFETs have a diode between the drain and source. When measuring the resistance between those terminals, it will be high in one direction and low in the opposite direction.

Some popular power MOSFETs and their characteristics are listed in Table 10.3. They commonly appear as shown in Fig. 10.21.

SILICON CONTROLLED RECTIFIERS

The silicon controlled rectifier (SCR) or is a four-layer device of the thyristor family. The symbol and structure are shown in Fig. 10.22. The SCR, as the symbol suggests, can be compared to a diode that, once triggered, conducts current in only one direction.

To trigger the SCR on, it is necessary to apply a positive voltage to the gate. This voltage biases the NPN transistor of the equivalent circuit to the saturated state. This results in feedback that makes the PNP transistor to go to the saturated state as well.

Even when the trigger voltage disappears, the transistors remain on due to feedback. To turn off the SCR, there are two options:

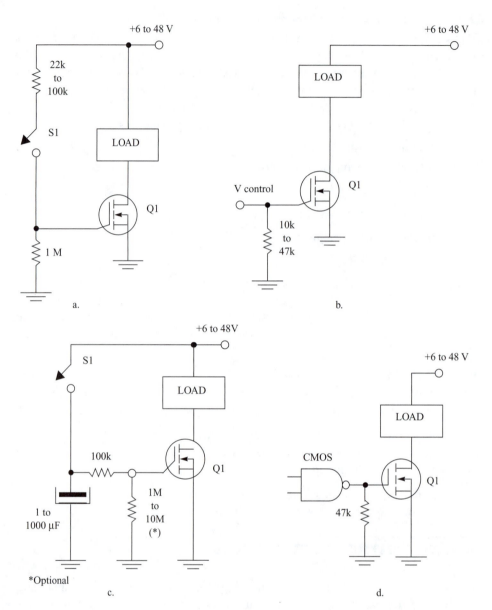

Figure 10.20

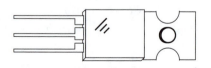

Figure 10.21

Table 10.3   Popular Power MOSFETS

	V_{ds} (V)	$R_{ds(on)}$ (Ω)	Ip (A)
IRF120/IRF520	100	0.3	8.0
IRF121/IRF521	60	0.3	8.0
IRF122/IRF522	100	0.4	7.0
IRF123/IRF523	60	0.4	7.0
IRF140/IRF540	100	0.085	27
IRF141/IRF541	60	0.085	27
ITF142/IRF542	100	0.11	24
IRF143/IRF543	60	0.11	24
IRF230/IRF630	200	0.4	9.0
IRF231/IRF631	150	0.4	9.0
IRF232/IRF632	200	0.6	8.0
IRF233/IRF633	150	0.6	8.0
IRF240/IRF640	200	0.18	18
IRF241/IRF641	150	0.18	18
IRF242/IRF642	200	0.22	16
IRF243/IRF643	150	0.22	16

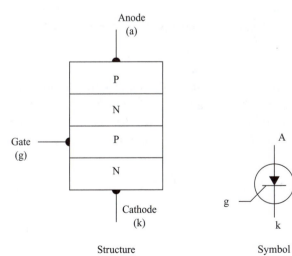

Structure   Symbol

Figure 10.22

1. Using a switch, short out the anode and cathode, making the voltage between them fall to zero. By pressing a momentary-contact switch, you cause the SCR turn off.

2. Turn off the circuit power. When the power is turned on again, the SCR remains off as long as no trigger voltage exists at the gate.

Common SCRs are very sensitive and can be triggered by currents as low as a few hundred microamperes. Examples include the TIC106, MCR106, IR106, and others. The trigger voltages for those devices range between 1 and 2 V. The primary currents (between anode and cathode) are typically in the range of 100 mA to more than 100 A.

SPECIFICATIONS

An SCR can be identified by the manufacturer's part number, which is generally printed on the package. From this part number, referring to an appropriate handbook, databook, or data sheet, it is possible to determine its electrical characteristics. When using any SCR, it is important to determine the characteristics described below.

Maximum Voltage between Anode and Cathode

This is the maximum voltage that can be applied to the SCR when it is in the off state. Typical values range from 50 to more than 1,000 V for common types. This specification may also be referred to as *repetitive peak off voltage* and abbreviated by V_{drm} or as *repetitive peak reverse voltage* and abbreviated by V_{rrm}.

Maximum Current

This refers to the maximum current that the SCR can carry when in the conductive state. Common SCRs, which the technician may find in appliances connected to the AC power line, are specified to control currents in a range from a fraction of an ampere to more than 50 A.

Dissipation Power

When an SCR is on, independent of the current it is conducting, the voltage between the anode and cathode is about 2 V. Multiplying this voltage times the current gives us the heat dissipation of the SCR. The maximum dissipation of an SCR is important in calculating the size and shape of the heatsink.

Holding Current

This is the lowest current that the SCR can conduct without turning off. This current is abbreviated by I_H.

Another important specification is the speed of the device (given by a value called *critical rate of rise of off-state voltage* and measured in volts per microsecond). This value tells us how quickly the voltage applied to the load increases when the SCR is turned on.

HOW TO USE SCRS

SCRs can be used both in AC and DC circuits. Just keep in mind that, in a DC circuit, the device remains on after the trigger signal is turned off. Figure 10.23 shows some practical circuits using SCRs, as enumerated below.

Figure 10.23a. AC power switch; SW1 is pressed to turn the circuit off

Figure 10.23b. DC power switch

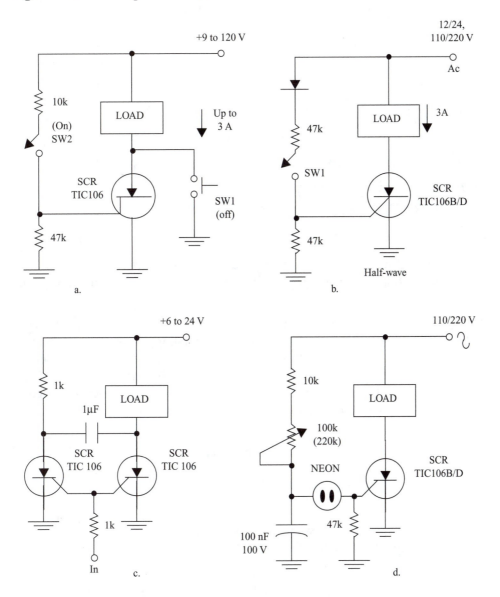

Figure l0.23 *(continues)*

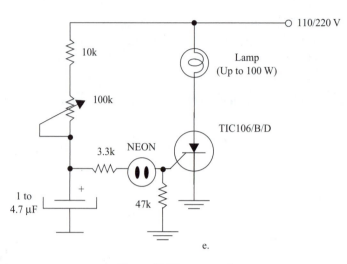

Figure 10.23   *continued*

Figure 10.23c.   Flip-flop using an SCR

Figure 10.23d.   AC dimmer

Figure 10.23e.   AC flasher

When an SCR is turned on, it requires a certain minimum current to keep it on. This current is called the *holding current* and is in the range of a few milliamperes for common devices.

Another important consideration when using SCRs in low-voltage circuits is that a voltage drop of about 2 V is produced when it is operating. In some of our applications, we must compensate for this voltage drop by an increase in power supply voltage. In high-voltage applications (e.g., AC power line connections), this voltage drop can be ignored. This voltage also determines the amount of power converted to heat by the device when in operation.

Formula:
The heat dissipated by an SCR can be computed as

$$Pd = V \times 1 \tag{10.4}$$

where Pd = dissipation (W)
        V = voltage drop across the SCR (typically 2 V)
        I = current across the SCR (A)

TEST METHODS
A simple SCR test using a multimeter is shown in Fig. 10.24, and its appearance is shown in Fig. 10.25. Common SCRs and their traits are shown in Table 10.4.

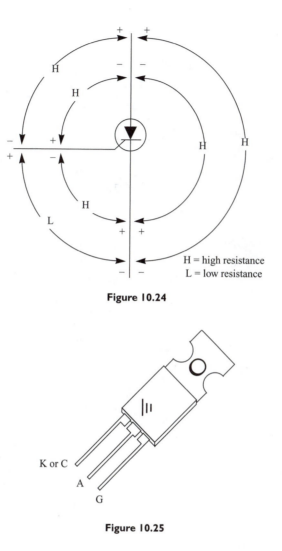

Figure 10.24

Figure 10.25

TRIACS

Triac make up another important device in thyristor family. Unlike SCRs, triacs are bidirectional and can conduct current in both directions. We can think of a triac as two SCRs connected in reverse parallel (anti-parallel) and placed inside a single three-terminal package as shown in Fig. 10.26.

The triac is suited to AC voltage control applications, meaning that it has a wide range of applications in mechatronic projects that are powered from the AC line. Like an SCR, a triac can be used as a power switch to control a large amount of alternating current from the main terminals (MT1 and MT2) using a low voltage applied to the gate.

Table 10.4 Common SCRS

Device	I_t (A)	V_{drm} (V)	$I_{gt(max)}$ (mA)	Device	I_t (A)	V_{drm} (V)	$I_{gt(max)}$ (mA)
T IC106Y	5	30	0.2	TIC116D	8	400	20
TIC106F	5	50	0.2	TIC116E	8	500	20
TIC106A	5	100	0.2	TIC116M	8	600	20
TIC106B	5	200	0.2	TIC126F	12	50	20
TIC106C	5	300	0.2	TIC126A	12	100	20
TIC106D	5	400	0.2	TIC126B	12	200	20
TIC116F	8	50	20	TIC126C	12	300	20
TIC116A	8	100	20	TIC126D	12	400	20
TIC116B	8	200	20	TIC126E	12	500	20
TIC116C	8	300	20	TIC126M	12	600	20

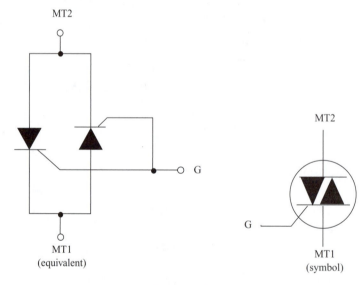

Figure 10.26

Unlike the SCRs, which need only a positive voltage applied to the gate to turn on, triacs can be triggered by either a positive or a negative gate signal, irrespective of the polarities of the main terminal voltages. From this we conclude that the device can be triggered in four modes, as follows:

I+ mode = MT2 positive and gate current positive
I– mode = MT2 positive and gate current negative
III+ mode = MT2 negative and gate current positive
III– mode = MT2 negative and gate current negative

The gate sensitivities in the I+ and III– modes are high; in the other modes, they are lower.

Common triacs need only a few milliamperes of current applied to the gate to turn on. Once set to the conductive state, high currents in the range from several to more than 1,000 A can be controlled.

SPECIFICATIONS

Triac packages are the same as used by high-power transistors, SCRs, and many other components. Note that they are basically high-power solid state switches and therefore must be mounted on heatsinks when handling high currents. Terminal identification (MT1, MT2, and G) is made using a device's data sheet or other manufacturer documents. Terminal configuration changes from type to type. A general rule is that MT1 is usually connected to the case or to the tab.

Triacs are identified by a part number printed on the package, as in the case of SCRs, transistors, and many other solid state devices. From the part number, you can refer to the appropriate data sheet or other documentation from the manufacturer to determine the electrical characteristics of the device. The reverse procedure is also valid; if you know what characteristics you need, the data sheet will give you the proper part number. The main specifications of interest in a triac are as described below.

Maximum Voltage

This is the maximum voltage that can be applied to the triac in the off state. This is also referred to as the *repetitive pulse reverse voltage* or *repetitive peak reverse voltage* (V_{rrm}). Typically, it will be in the range of 50 to more than 1,000 V for common types.

Maximum Current

The maximum current that can be controlled by the device is called the *continuous on-state current* or *average on-state current* and can be in the range of 1 A to more than 100 A for common types.

Trigger Current and Voltage

The current that triggers the triac on is called the *gate trigger current* and abbreviated by I_{gt}. Typically, this current is in the range of several microamperes to 100 mA, depending on the size of the device. The trigger voltage is also indicated as V_{gt} and is typically in the range of 0.8 to 1.5 V for common types.

Dissipation Power

When a triac conducts, the voltage across it is typically between 1.7 and 2.2 V. This voltage times the controlled current, or load current, gives the dissipation power of the device. The maximum allowable dissipation depends on the device.

HOW TO USE TRIACS

Figure 10.27 shows two basic circuit applications of a triac for controlling an AC motor, solenoid, or resistive load.

Figure 10.27a.   Speed control or dimmer

Figure 10.27a.   Power switch

The load is connected in series with main terminal 2 (MT2), and main terminal 1 is set to ground.

There are four ways to trigger the triac, as described previously (see p. 220). In basic applications, the first-mentioned trigger mode is preferred, as it provides more sensitivity.

When turned on, a voltage drop of about 2 V can be observed in the device. This voltage drop determines the amount of heat produced by the device (voltage drop × current = dissipation power).

Formula:

Power generated by a triac when in the full conduction state is

$$Pd = 2 \times I \tag{10.5}$$

where Pd = power (W)

I = rms current across the triac (A)

Obs.: "2" is the typical voltage drop across a triac in the fully conductive state.

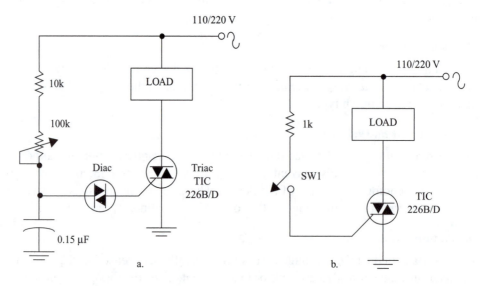

Figure 10.27

ELECTROMAGNETIC INTERFERENCE

Triacs and SCRs are high-speed power switches. Their turn-on and turn-off times are very short (only few microseconds), and this can cause electromagnetic interference (EMI). EMI (also referred to as radio frequency interference, or RFI) is interference caused by the undesirable radio frequency signals produced by devices like thyristors. These signals can adversely affect the operation of many sensitive electronic instruments. Radio control systems as used in many mechatronic projects can be disrupted by interference from triacs and SCRs operating nearby.

When an SCR or triac is used in a circuit for power control, the high-speed current changes in the load generate a series of harmonically related radio-frequency signals. The magnitude of the fundamental signal is proportional to the magnitude of the controlled current, and, in many cases, it can be great enough to cause interference in AM radios and many other circuits that use low- and medium-frequency radio signals in their operation.

In particular, the interference level is higher when the SCR or triac controls inductive loads as motors, solenoids, electromagnets, and so on. The snap action of the triac or SCR causes current oscillations in the circuit, generating a strong electromagnetic field.

A protective circuit that minimizes interference is shown in Fig. 10.28. The capacitor and the resistor form a "damping" circuit that reduces current oscillation and electromagnetic interference. Examples of the EMI process include the noise in an AM radio receiver and the static displayed on a TV screen when a circuit using triacs or SCRs is turned on.

Interference can enter radio receivers, TVs, and other sensitive devices in two ways:

1. It can be picked up from radio frequency emissions in the atmosphere (radiated emissions). In this case, the undesirable signal is generally very weak and causes problems only if a radio or other sensitive piece of equipment is placed

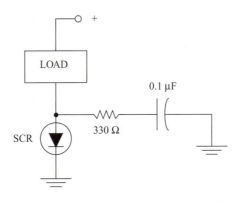

Figure 10.28

near the interference source. It is also concentrated in the low band of the radio spectrum, as shown in Fig. 10.29; the interfering signal is more powerful in the low-frequency (LF) band than in the medium-wave (MW) band.

Radio receivers operating at higher frequencies, such as FM receivers and VHF receivers (e.g., remote controls for garage doors and robots, and cellular telephones), aren't affected much by this interference. This is due to the lower power concentrated in this band as well as their basic operating principles.

2. In the second case, the signal is conducted into the victim equipment via its own AC power line. To avoid this type of interference, it is sufficient to use an L-C filter such as shown in Fig. 10.30.

Common triacs and their characteristics are listed in Table 10.5, and the appearance of a typical device is shown in Fig.10.31.

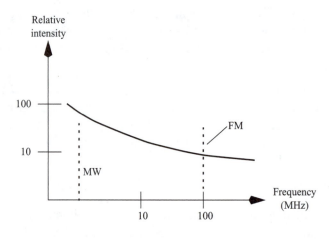

Figure 10.29

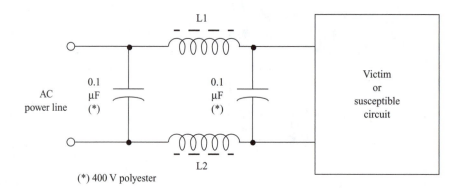

Figure 10.30

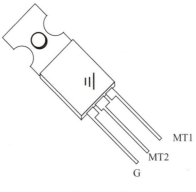

Figure 10.31

Table 10.5   Common Triacs and Characteristics

Type	$I_{t(rms)}$ (A)	V_{drm} (V)	$I_{gt\,max}$ (mA) I, II, III
TIC216A	6	100	5
TIC216B	6	200	5
TIC216D	6	400	5
TIC226B	8	200	50
TIC226D	8	400	50
TIC236B	12	200	50
TIC236D	12	400	50
TIC246B	16	200	50
TIC256D	16	400	50

ISOLATED-GATE BIPOLAR TRANSISTORS

The isolated-fate bipolar transistor (IGBT) is a semiconductor device that combines the advantages of a bipolar transistor and a power MOSFET. IGBTs are transistors in which the main current or controlled current flows between a collector and an emitter as in a bipolar transistor; however, it is controlled by a gate voltage as in an FET. The advantage is that the device presents a very high input impedance along with the control characteristics of a bipolar transistor.

IGBTs are used for the same functions as power bipolar transistors and power MOSFETs and, in some cases, with significant advantages. They are used in applications where we need to drive a high-current load (e.g., solenoids, lamps, motors, etc.). Figure 10.32 shows the circuit symbol for an IGBT.

As in the case of other transistors, we can't identify an IGBT by its appearance only, as they are found in the same package styles as bipolar transistors and power

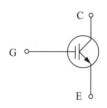

Figure 10.32

MOSFETs. We must have a part number or a schematic diagram of the equipment in which it is used.

HOW IGBTS WORK

IGBTs are used in the same manner as bipolar power transistor and power MOS-FETs. In basic applications, they are wired as shown in Fig. 10.33. Applying a positive voltage of several volts to the gate element will cause a high current to flow across the device (i.e., between collector and emitter) and the load.

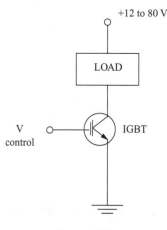

Figure 10.33

SPECIFICATIONS

IGBTs can be identified by part number, from which the electrical characteristics can be determined. The main characteristics are as follows.

1. *Maximum voltage between collector and emitter.* This is the maximum voltage that can be applied to the device, abbreviated by $V_{ce(max)}$.

2. *Maximum collector current.* This is the maximum current that can be conducted by the device, abbreviated by $I_{c(max)}$.

3. *Dissipation power.* This is the maximum power that the device can convert into heat and dissipate into the ambient. It is given in watts and abbreviated by $P_{d(max)}$ or Pd.

Common IGBTs are listed in Table 10.6, and Fig. 10.34 shows the circuit symbol and pinouts.

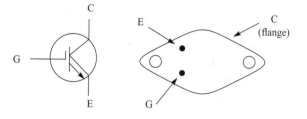

Figure 10.34

Table 10.6   Common IGBTS

	2N6975/2N6977	2N6976/2N6978[*]
Collector-emitter voltage (max)	400 V	500 V
Collector-gate voltage (max)	400 V	500 V
Gate-emitter voltage (max)	5 V	5 V
Collector current (max)	5 A	5 A
Power dissipation (max)	100 W	100 W

[*] Harris Semiconductor devices.

POWER INTEGRATED CIRCUITS

The integrated circuit should not be considered as a single component but as a collection of components manufactured on a single chip and installed inside an unique package. The basic concept came from asking the following questions:

1. Why should we have to manufacture all circuit components individually and then assemble them to achieve the desired configuration?

2. Why not manufacture a circuit, with the components already connected, on a single silicon chip to perform basic desired circuit functions, using a single fabrication process?

The answer to these questions emerged in the form of the integrated circuit (IC).

The IC consists of a tiny silicon chip that includes a predetermined number of components (e.g., transistors, resistors, capacitors, diodes, etc.) that are interconnected by conductive paths that act as connections. Components and conductive paths (also called *traces*) are formed by a complex process that includes diffusion, photolithography, and etching.

The number of components, their positions on the chip, and the interconnects are designed to form a complete functional circuit such as an amplifier, motor control, voltage regulator, or even a computer. On a single chip, it is possible place circuits consisting of only a few components that perform simple functions (e.g., a collection of matched transistors), but it can also contain more than 10,000,000 components as in the case of modern microprocessors used in personal computers. Even a complex integrated circuit with millions of transistors may have a footprint as small as a few square millimeters. The principal advantage of the IC is that the circuit can be manufactured as a nearly complete electronic device, in some cases needing only a few complementary support devices to form a complete piece of electronic equipment.

The integration process does have some limits. Some components cannot be integrated on a silicon chip. For example, high-value capacitors and inductors cannot be easily incorporated in integrated circuits. When designing a complex IC, we must allow for the fact that such components must be external, and the traditional *discrete* type of device must be used. Therefore, although an IC can allow us to simplify our projects, it cannot always form the complete circuit or equipment; it needs to be complemented with external components such as capacitors and inductors.

Since integrated circuits can contain practically any configuration and any number of components, they are available in a wide variety of types and functions. Figure 10.35 shows common symbols and package appearances of ICs. In most cases, an identification symbol or part number is printed on or inside an IC to allow the technician to identify it.

The number of terminals (leads) on an IC varies according to its function. The most common types have 8 to 40 or more leads. Some ICs for special applications may have more than 200 leads.

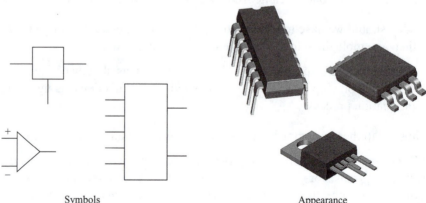

Symbols   Appearance

Figure 10.35

In Fig. 10.35, we show some of the most commonly used IC types, found in a range of electronic equipment. Typical packaging styles include the dual in-line package (DIP), single in-line package (SIP, not shown), small-outline IC (SOIC), and TO-220. Note that the terminals are often identified in relation to a mark or depression at one end of the package. Terminal identification is important when installing the IC and when making voltage measurements to troubleshoot equipment.

Some integrated circuits that perform high-power functions (e.g., motor controls, voltage regulators, audio amplifiers) are equipped with tabs or other resources that allow them to be mounted on heatsinks.

SPECIFICATIONS

ICs are identified by part number. Only with the specific part number is it possible to know what an IC is and what it does. The mechatronics professional or student who wants to work with electronic components does not need to have complete handbooks or manuals for all integrated circuits—which would be impossible. ICs can be divided into primary groups as described below.

Basic Functions

Some integrated circuits are simple enough and useful enough that they can be employed in a wide variety of applications. These *general-purpose* circuits act as basic building blocks in many applications and are often present in mechatronic projects. These are cheap and easy to use, which puts them in a different class from *special-function* ICs, which can require an investment of thousands of dollars. General-purpose ICs can be subdivided according to their operational mode.

Analog vs. Digital

Analog ICs work with DC and analog signals such as audio, video, AC, and so on. We define an *analog* value as one that can assume any value between two limits. An AC current is an analog quantity, as its amplitude can assume any level between zero and the maximum. Conversely, a digital quantity can assume only specific values between two limits. For example, a four-bit "nibble" (half a byte) can assume only the integer values between 0000 and 1111.

Within the group of analog ICs, we find such devices as voltage regulators, operational amplifiers, comparators, audio amplifiers, video amplifiers, timers, PLLs, and others. It is relevant to give a brief description of these ICs, as they are quite commonly used in many electrical systems, particularly in appliances that are powered from the AC line.

VOLTAGE REGULATORS

The purpose of a voltage regulator is to maintain a constant circuit voltage. Many equipment power supplies use this type of IC to provide a regulated output voltage. In modern appliances, a three-terminal, 1-A voltage regulator is common. Examples include the 78XX and 79XX series. In this designation, the XX is replaced with a figure that indicates the output voltage. For example, the 7806 provides a 6-V output under 1 A (maximum) current.

Another type of voltage regulator is the switch-mode power supply (SMPS) as commonly used in computers, video monitors, robots, and many other modern appliances. These regulators differ from linear power supplies in that they act as oscillators that produce a voltage that depends on the frequency and pulse width of the output signal. See Chapter 11 for more information about these devices; they are also used in power supplies for mechatronic devices.

OPERATIONAL AMPLIFIERS

Operational amplifiers (OAs or opamps) are general-purpose amplifiers that were originally designed to perform mathematical operations in analog computers. An operational amplifier has a non-inverting input represented by a plus (+) and an inverting input indicated by a minus (−). The signals applied to the inverting input appear at the output with the phase shifted, as shown in Fig. 10.36.

The ideal operational amplifier has an infinite input impedance and a null output impedance. In the real world, they have a very high input impedance of many megohms and a very low output impedance (several to hundreds of ohms).

The gain of an operational amplifier (i.e., the number of the times the output voltage is higher than the applied input voltage) can be controlled by a feedback circuit as shown in Fig. 10.37. Higher gain is achieved when the feedback is infinite and typically ranges from 10,000 to 1,000,000 in common types. The lowest gain is achieved when the feedback loop is a 0-Ω resistance, and it equals 1. The gain of an operational amplifier can also be expressed in decibels (dB).

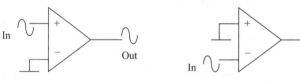

Non-inverting   Inverting

Figure 10.36

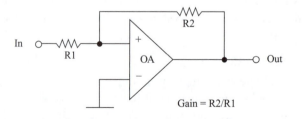

Gain = R2/R1

Figure 10.37

The configuration with a voltage gain of 1 is called a *voltage follower*. In a voltage follower, the output voltage is the same as the input voltage. Even so, the difference between the input and output impedance makes the circuit an amplifier with a very high current gain.

Operational amplifiers can be found in many types and sizes and with a wide range of electrical characteristics [i.e., input and output impedances, power supply voltages, gains, type of transistors found inside (FETs or bipolars), operational frequency range, and so on]. Figure 10.38 shows a symbol that has been adopted to represent operational amplifiers. The packages are the same as used for any other IC.

Note that some ICs contain two or more opamps. In some cases, we can also find terminals for the connection of additional components. The external components add special characteristics to the circuit such as frequency compensation and others.

Opamps are designated by part number according a code adopted by the manufacturer. Normally, a group of letters is used to identify the manufacturer (e.g., MC = Motorola, SN = Texas Instruments, LM = National Semiconductor, and so on).

By cross-referencing the part number with data sheets or handbooks, it is possible to determine the electrical characteristics of each device. Primary characteristics are as follows:

1. *Open-loop gain.* The gain is determined by the internal circuit and is specified as the maximum gain (or open-loop gain) when no feedback is present. The values can range from 1,000 to 1,000,000 in common types. The gain of an OA is often abbreviated by the letter G.

2. *Power supply voltage range.* This specification indicates the voltage that the opamp needs for correct operation. It can range from 1.5 to more than 40 V.
   At this point, we should observe that, in many applications, the opamp needs a symmetric or dual power supply. This is a power supply in which both

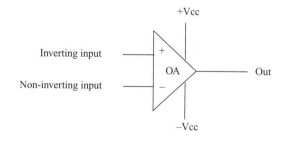

Figure 10.38

a positive and a negative voltage (referenced to the 0-V ground) must be generated as shown in Fig. 10.39.

In some documents, data sheets, and diagrams it is common to refer to the power supply voltage for an OA as 6-0-6 V, +6, –6, 0 V, +6/–6/0 V, or 6+6 V if it needs a symmetric supply such as the one shown.

3. *Gain × bandwidth.*   The gain of an operational amplifier falls as the signal frequency increases. The frequency band is limited by the point at which the gain falls to 1, which determines the gain bandwidth product. Common operational amplifiers can operate at frequencies ranging from 1 or 2 MHz up to several hundred megahertz.

4. *CMRR.*   If two signals with the same amplitude and phase are applied to an opamp, one to the inverting input and the other to the non-inverting input, they will be cancelled, and the output will be zero. In practice, due to small variations in the internal components, this is not possible; some part of the signal will not be cancelled and will appear at the output. The efficiency of the OA when cancelling a signal in this operational mode is given by a specification called *common mode rejection ratio* (CMRR). This specification, which should be as high as possible, is indicated in decibels (dB). Common types have a CMMR as high as 90 dB or more.

Many appliances use operational amplifiers in the integrated circuit form. They normally will be common types such as the 741, CA3140, TL080, TL081, and many others. These can be found in any electronic component shop.

Special operational amplifiers can be salvaged from circuits designed to drive motors, solenoids, and other devices, but, in general, you can obtain what you need from dealers by referencing their part numbers.

In many circuits, the basic function of an operational amplifier is the amplification of signals that come in from sensors, transducers, or other sources, boosting them to drive powerful output stages or circuits that demand higher-amplitude signals.

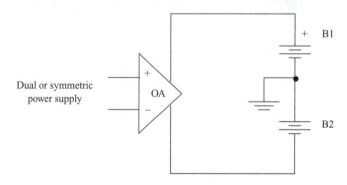

Figure 10.39

Figure 10.40 shows some practical opamp circuits that are suitable for mechatronics applications in mechatronics. They are based in the most popular device, the 741, but equivalents can be substituted. Shown in the figure are the following:

Figure 10.40a. Voltage follower (gain = 1)

Figure 10.40b. Gain 100

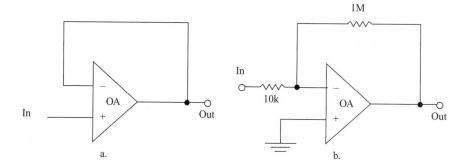

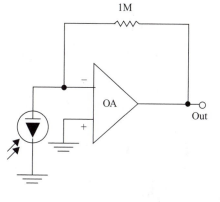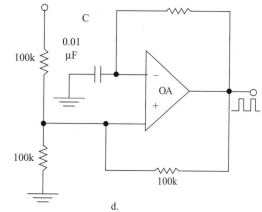

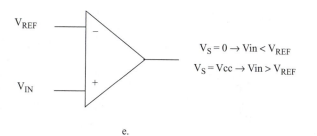

$$V_S = 0 \rightarrow Vin < V_{REF}$$
$$V_S = Vcc \rightarrow Vin > V_{REF}$$

Figure 10.40

Figure 10.40c.   Amplifier for transducer

Figure 10.40d.   Low-frequency oscillator

Figure 10.40e.   Voltage comparator

Test Methods

In many applications, an opamp can be tested by measuring the voltages at its pins. If the voltages are incorrect, and the components connected to it are good, the problem is certainly with the opamp. An opamp can also be tested using a special probe circuit.

OPERATIONAL AMPLIFIER GAIN

The gain of an operational amplifier in the configuration given by Fig. 10.41 is calculated as

$$G = \frac{R_1}{R_2} \tag{10.6}$$

where
$G =$ gain (V/V)
R_1 and $R_2 =$ resistances in ohms

Table 10.7 shows conversions of voltage ratio to dB, and Table 10.8 lists some common opamps and their characteristics.

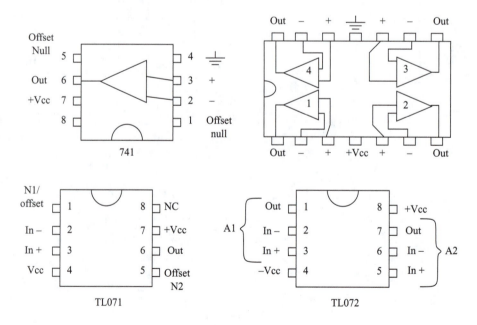

Figure 10.41

Table 10.7 Voltage Ratio to dB Conversions

Voltage ratio	dB	Voltage ratio	dB
0.001	−60	100	+40
0.01	−40	316	+50
0.1	−20	1,000	+60
1	0	3160	+70
2	+6.02	10,000	+80
3	+9.54	31,600	+90
4	+12.04	100,000	+100
5	+13.98	316,000	+110
10	+20	1,000,000	+120
31.6	+30		

Table 10.8 Common Operational Amplifiers

	Gain (tip)	Supply voltage	Input impedance (tip)	Output impedance (tip)
741[*]	200,000	18 + 18 V	2 MΩ	75 Ω
LM124/224/324	100,000	16 + 16 V	−	−
TL071/TL072	150,000	18 + 18 V	10^{12} Ω	−

[*] LM741, MC1741, etc.

AUDIO AMPLIFIERS

Complete audio amplifiers with output powers ranging from several milliwatts to more than 100 W can be found in the form of linear integrated circuits. These ICs include several stages of amplification, as they contain a preamplifier, a driver, and output stages, all interconnected to make up a complete audio amplifier.

The high-value capacitors required by many of the stages are connected to the circuit via external pins. In many cases, manufacturers can build such a circuit using a minimum of external capacitors.

Many audio amplifiers allow us to control the gain or the frequency response using some external components. In the simplest types, we have an input where a volume control is connected, an output where a loudspeaker or headphones/earphones are connected, power supply pins, and whatever connections are required for external capacitors (coupling and decoupling) and resistors (for gain control and so forth).

The basic function of an audio amplifier is to drive a loudspeaker, earphone, or headphone with a signal that results in audible sound. However, ICs intended for this task can also be used as high-power operational amplifiers to drive DC motors or the transformers of inverters. These amplifiers have many applications in mechatronic projects such as adding sound effects and driving motors and inverters. Many manufacturers offer devices that contain two amplifiers in one package, making it easier to incorporate stereo sound in our equipment.

Figure 10.42 shows the basic symbol that is used to represent an audio amplifier. The devices may be packaged as SIPs, DIPs, or other types, depending on the output power.

ICs in this group are also identified by the manufacturer's part number. The electrical specifications of an audio amplifier IC are the same as those of any other audio amplifier, as discussed below.

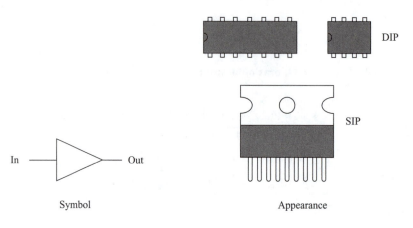

Figure 10.42

Output Power

This is the amount of power that the circuit can deliver to a loudspeaker. It is indicated in watts (W) and in some cases abbreviated by P_o or P_{out}.

Power Supply Voltage and Current

This specification is important when designing a power supply for the circuit. It is given in volts.

Output Impedance

The characteristics of the loudspeakers or earphones that will be connected to the output of any audio amplifier are very important, as we do not want to overload the amplifier. The impedance is given in ohms and typically is between 2 and 16 Ω for loudspeaker drivers. The impedance may be higher for earphone/headphone amplifiers.

Input Impedance

This specification is given in ohms.

Gain

Gain is defined as the ratio of the output voltage to the input voltage. For example, with a tenfold increase in power (output-to-input ratio of 10:1), the amplifier gain is 10. This specification is important because, along with the input impedance, it is used to determine the sensitivity of an amplifier.

Operational Curves and Other Specifications

The performance of an audio amplifier versus signal frequency is usually indicated by curves on a graph. These response curves illustrated the fidelity of an amplifier when working with signals in the audio spectrum.

These specifications are important when designing a project. For the technician who just needs to find a replacement part, the part number is sufficient. It is important to remember that amplifiers with equivalent power specifications may not be interchangeable, as their package design and pinouts may be different.

Audio amplifiers are found in many common AC-powered appliances and embedded in various electrical circuits, and some are powered from batteries (e.g., in talking robots and other mechatronic devices).

Testing Amplifier ICs

The best way to test a common audio integrated circuit is to measure the voltages at its pins. It is recommended that you perform these measurements only after you have verified that associated components (external capacitors, resistors, and so on) are functional. To test the IC, you must have some idea about the what the voltages should be, and it is best to have a schematic diagram that indicates these values.

TIMERS

Another important family of linear integrated circuits consists of timers. These are circuits that can be used to generate a signal or to count by time intervals. Basically, they can be used in two operational modes: *monostable* and *bistable*.

In the monostable mode, after being triggered, the circuit produces a signal over (or after) a particular time interval to trigger or control something as suggested by Fig. 10.43. After this interval, the circuit returns to its previous (stable) state. In bistable operation, the circuit can remain stable in either of the two states.

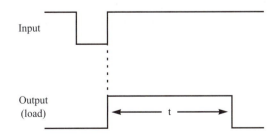

Figure 10.43

There is also the *astable* or unstable mode of operation in which the circuit generates a square signal at a frequency that depends on some external components, normally an R-C network, as shown in Fig. 10.44. In this mode, the device is not stable in either state.

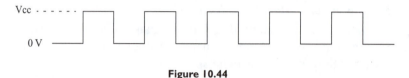

Figure 10.44

THE 555

The most common of the commercially available timers is certainly the 555, which can be found with designations indicating the various manufacturers, such as NE555, TLC7555, LM555, MC555, and so on. This integrated circuit can operate in both astable and monostable modes, depending on the way some external components are connected (see Fig. 10.45).

The bipolar version (in which the internal transistors are bipolar types) can provide time intervals up to 1 hr and oscillation frequencies up to 500 kHz. A version using JFET transistors (TLC7555) can provide even longer time intervals and higher frequencies, with lower current consumption.

Specifications

Timers packaged as integrated circuits are specified by a part number, as are other ICs. The electrical characteristics can be found in the manufacturer's documenta-

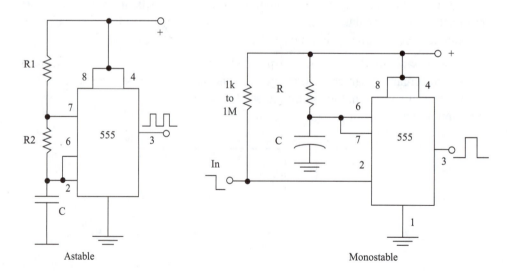

Figure 10.45

tion or the derived from schematics of the equipment in which they are used. The main electrical specifications are described below.

Power Supply Voltage

This specifies the power supply voltage range that is suitable for the circuit.

Time Interval or Frequency

In practical applications, external components (R and C) determine the timer's operational frequency. The data sheet should provide a formula or graph that allows the designer to choose the correct value of these components for a particular application.

Output Current

This is an important consideration, as it tells you what you can control with the IC and how to do it. In some cases, a transistor stage must be used to control a device that has high current consumption.

Operational Modes

Many timers are capable of different operational modes. These are described in the specifications, which also provide data about how the external components must be connected to obtain a particular mode of operation.

Testing the 555

As with many other ICs, there is no generic test procedure. All ICs are different, and test procedures vary according to the device. The best way to determine if the IC is functioning properly is to check the voltage at each pin and compare it to data sheet specifications or schematic values. This approach provides accurate readings only if other components in the circuit are good.

ASTABLE 555—USEFUL FORMULAS

The following formulas can be helpful in using the astable 555 (refer again to Fig. 10.45).

Charge time of C (output high)

$$Th = 0.693 \times Cx(R_1 + R_2) \tag{10.7}$$

where T_h = time interval of the output high in seconds (s)

R_1 and R_2 = resistances in ohms (Ω)

C = capacitance in farads (C)

Discharge time of C (output low)

$$T_L = 0.693 \times R_2 \times C \tag{10.8}$$

where T_L = time interval of the output low in seconds (s)

R_2 = resistance in ohms (Ω)

C = capacitance in farads (F)

Period

$$T = 0.693 \times (R_1 + 2 \times R_2) \times C \qquad (10.9)$$

where T = period in seconds (s)

R_1 and R_2 = resistances in ohms (Ω)

C = capacitance in farads (F)

Frequency

$$f = \frac{1.44}{(R_1 + 2 \times R_2) \times C} \qquad (10.10)$$

where f = frequency in hertz (Hz)

R_1 and R_2 = resistances in ohms (Ω)

C = capacitance in farads (F)

Duty cycle

$$Dc = \frac{T_h}{T_L}$$

$$Dc = \frac{R_1 + R_2}{R_2} \qquad (10.11)$$

where Dc = duty cycle (0 to 1)

T_h = time the output is high in seconds (s)

T_L = time the output is low in seconds (s)

R_1 and R_2 = resistances in ohms (Ω)

Duty cycle as a percentage

$$Dc(\%) = Dc \times 100 \qquad (10.12)$$

where $Dc(\%)$ = the duty cycle in percents (%)

Dc = the duty cycle as calculated using Eq. (10.11)

Recommended limit values for the astable 555 are shown in Table 10.9.

Table 10.9   Recommended Limit Values for the Astable 555

Component	Limit value	Component	Limit value
R1 + R2	max: 3 MΩ	C	max 2200 μF*
R1	min 1 kΩ	f	max 1 MHz
R2	min 1 kΩ	I_{out} (drain or source)	200 mA
C	min 500 pF	Vcc	18 V

* Depending on leakage.

MONOSTABLE 555—USEFUL FORMULA

When connected as a monostable multivibrator, the 555 needs an external trigger command applied to the trigger input (pin 2) to initiate the action. This is normally done by leaving the trigger input positive and momentarily setting it to ground. In this manner, the output goes to the high logic level (positive) by a time interval that can be calculated using the following formula (see Fig. 10.45).

Time on

$$T = 1.1 \times R \times C \tag{10.13}$$

where T = period or time on in seconds (s)
R = resistance in ohms (Ω)
C = capacitance in farads (F)

Recommended limit values for the monostable 555 are shown in Table 10.10.

Table 10.10   Recommended Limit Values for the Monostable 555

Parameter/component	Limit value	Parameter/component	Limit value
R max	3 MΩ	Tr* (max)	1/4 T
R min	1 k Ω	Iout† (drain or source)	200 mA
C max	2000 μF‡	Vcc§	18 V
C min	500 pF		

* Trigger pulse duration in seconds.
† Maximum output current.
‡ Depending on leakage.
§ Power supply voltage.

PHASE LOCKED LOOPS

Phase locked loops (PLLs) have many important applications in mechatronics. A PLL is a circuit that can recognize a signal at a specific frequency and act on a relay or other device. If the signal is frequency modulated, the PLL can also be used to extract the modulation signal. Many types of PLLs exist in IC form.

PLLs are low-power circuits, but they can control high-power circuits that drive motors, solenoids, relays, and other loads. As with any other IC, PLLs are designated by part number.

From the part number, we can determine the device's electrical specifications, which are important when designing with a PLL or even when finding a replacement part. The main specifications are as described below.

1. *Power supply voltage range.*   This specifies the device's functional voltage range.

2. *Frequency range.*   This is a very important specification, as it indicates the highest signal frequency that can be recognized by a PLL, or generated by it when operating as an oscillator. Common general-purpose PLLs such as the 567 can operate at frequencies up to 500 kHz.

3. *Sensitivity.*   This refers to the minimum signal amplitude that can be recognized by a PLL.

4. *Output current.*   This specification indicates the amount of current the device can deliver or drain when driving a load.

In many modern applications, PLLs are included in more complex ICs. But there are also devices that offer the PLL as a single chip. Some common examples are the 4046, LM567, and other popular devices.

The following applications are indicative of how useful PLLs can be:

1. *Remote control.*   A PLL can be used to recognize a tone sent by a transmitter (in the radio frequency or infrared band) and act on another device (a garage door opener motor or other electric or electronic appliance).

2. *Intercoms.*   One type of intercom uses the AC power line to carry audio signals. The sounds picked up by a microphone are used to modulate a circuit, which produces an FM signal that is applied to the AC power line. A PLL in the receiver detects this signal and sends it to an amplifier.

3. *Alarms and sensors.*   A tone is produced by a circuit and applied to a PLL via sensors. If any sensor is activated, the PLL detects the absence of the tone and triggers an alarm.

Test Procedures

The same test procedure recommended for ICs is valid here: measure the voltages at the pins and compare the readings with expected values.

Practical Circuits

Two practical circuits using the LM567 are shown in Fig. 10.46. They are as described below.

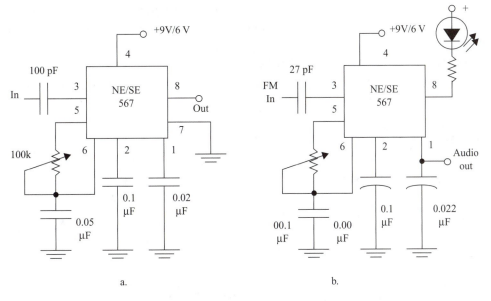

Figure 10.46

Figure 10.46a. Tone detector—this circuit can be used to recognize a tone between 100 Hz and 200 kHz, depending on the adjustment of P1. When the tone is recognized, the output goes to the low logic level, driving the LED or relay.

Figure 10.46b. FM demodulator—this circuit can extract the low-frequency component of a high-frequency FM signal.

Characteristics of the NE/LM567 are shown in Table 10.11.

Table 10.11   Characteristics of the NE/LM567

Frequency range	0.1 to 500 kHz
Maximum operating voltage	10 V
Input resistance (tip)	20 kΩ
Smallest detectable voltage	20 mVrms
Operating voltage range	4.5 to 9 V
Highest center frequency	500 kHz
Output voltage (LOW)—$I_L = 100$ mA	0.6 V (tip)

OTHER LINEAR IC FUNCTIONS

Many other functions can be performed by linear ICs. In addition to those previously described, the reader can obtain ICs containing video amplifiers, comparators, voltage references, radio receivers, and others. There is no specific symbol to indicate these functional variations, and ICs of this type look the same as other ICs.

DIGITAL INTEGRATED CIRCUITS

Digital integrated circuits form a special group of devices that have a wide range of applications in modern electronics. Digital ICs are discussed in Chapter 16.

UNIJUNCTION TRANSISTORS

The unijunction transistor (UJT) is a device that provides a negative input resistance. This characteristic makes it useful as an oscillator in low-frequency applications. Figure 10.47 shows the symbol and structure of this device.

The UJT can be used as a relaxation oscillator in the configuration shown in Fig. 10.48. This circuit can be used to produce sawtooth waves or pulses in the frequency range between a fraction of a hertz and 100 kHz. The most popular of the unijunction transistors is the 2N2646.

FREQUENCY AND PERIOD OF THE UNIJUNCTION OSCILLATOR

The frequency and period of a unijunction oscillator can be approximated as follows:

$$f = 1/T \tag{10.14}$$

$$T = R \times C \tag{10.15}$$

where f = frequency (Hz)
T = period (s)
R = resistance (Ω)
C = capacitance (F)

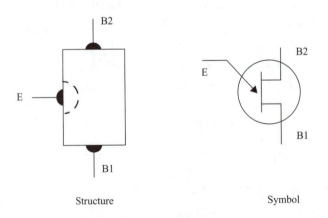

Structure   Symbol

Figure 10.47

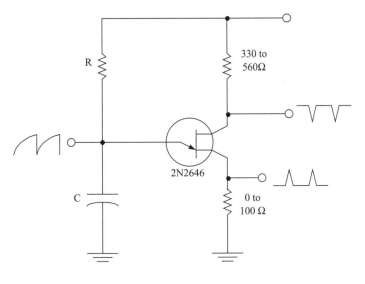

Figure 10.48

UJT APPLICATIONS

Figure 10.49 shows some practical circuits using the UJT. These are suitable for mechatronic applications. Figure 10.49a shows a siren, Fig. 10.49b is an LED blinker, and Fig. 10.49c shows an alarm circuit.

NEON LAMPS

The neon lamp is not really a solid-state device, of course. But this device, because of its negative resistance characteristic, is useful in many circuits as a trigger element. Figure 10.50 shows the symbol and appearance of a neon lamp.

The neon lamp passes from the off state to the on state, conducting a current when the voltage across its terminals rises to about 80 V. In many applications, the neon lamp can replace diacs and SBS triggering thyristors such as triacs and SCRs.

NEON LAMP APPLICATIONS

Figure 10.51 illustrates a simple low-power flasher that uses a neon lamp.

DIACS

The diac is a trigger device for triacs. This semiconductor device passes quickly from the off state to the on state when the voltage across it rises to a certain value (about 27 V for common types). Because the diac triggers on with negative or positive voltages, it can be used to trigger triacs as shown in Fig. 10.52.

In some cases, the diac and the triac are mounted inside the same package. The device formed by these two elements is called "quadrac."

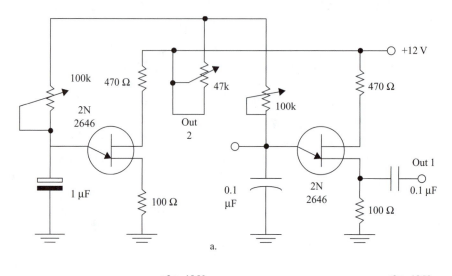

Out 2

Out 1

a.

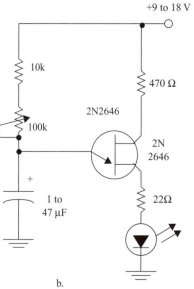

b.

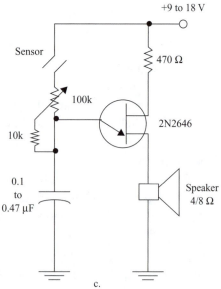

c.

Figure 10.49

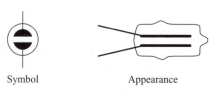

Symbol

Appearance

Figure 10.50

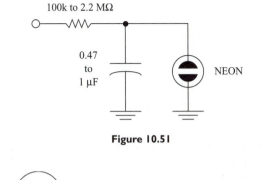

100k to 2.2 MΩ

0.47
to
1 μF

NEON

Figure 10.51

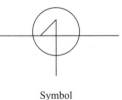

Symbol

Appearance

Figure 10.52

SILICON UNILATERAL SWITCHES

The silicon unilateral switch (SUS), like the neon lamp, displays a negative resistance characteristic that makes it useful as trigger device for SCRs. Figure 10.53 shows the symbol for this device.

Symbol

Figure 10.53

SILICON BILATERAL SWITCHES

The silicon bilateral switch (SBS) is another trigger device for circuits using a thyristor. The symbol and appearance of this component are shown in Fig. 10.54. The difference between this device and a diac is that the trigger voltage can be governed by the control electrode.

VACUUM TUBES

Although ICs and transistors have replaced tubes in most modern applications, much equipment is still in use that is based on these components. To understand the operation and use of tubes, we can start with the simplest type: the diode.

Symbol   Appearance

Figure 10.54

Inside a glass tube, we have some electrodes in a vacuum. In the case of a diode, a tungsten filament is placed near an electrode called the *cathode*. Nearby the cathode (but not touching it) is another electrode called the *anode*. To make the tube work, a positive voltage is applied to the anode, the cathode is connected to a negative voltage, and the filament is heated by a low-voltage source as shown in Fig. 10.55.

As the filament raises the temperature of the cathode, electrons are emitted, causing a current flow between the cathode and anode. If the anode is negative and the cathode positive, no current can flow through the tube. In a tube like this, the current can flow only in one direction as in a semiconductor diode.

If we now place a metallic grid between the anode and the cathode, as shown in Fig. 10.56, we can observe new electrical properties in the device. A voltage applied to this grid can control the electron flow between the cathode and anode. If the grid is made negative, for instance, the electrons are repulsed, and no current can flow. On other hand, if the grid is made positive, the electrons are attracted and accelerate to the anode.

The amount of current between the anode and cathode can be precisely controlled by the voltage applied to the grid. This device, called a triode, can be thought of as "equivalent" to the transistor, and it is used for the same purposes: as a switch or amplifier.

Newer tubes were derived from the triode configuration by adding new internal elements. The tetrode (four elements), pentode (five elements), hexode (six elements),

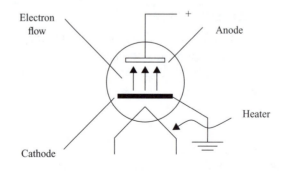

Figure 10.55

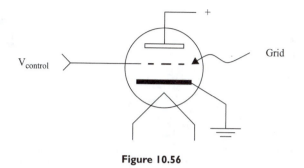

Figure 10.56

and heptode (seven elements) are examples of tubes that can be found such old equipment as radio receivers, audio amplifiers, and TVs.

Although tubes are very efficient in some applications and are even used in some modern equipment, they have some disadvantages when compared with equivalent modern solid-state devices such as transistors. For one thing, tubes need to be heated to operate, and this implies the consumption of large amounts of energy. Another problem is that all the heat is usually transferred to the ambient. Tubes also need high voltage for their operation. Although common solid-state transistors operate with low voltages, ranging from 1 to 30 V in typical applications, tubes, in the same applications, typically need voltages in the range of 150 to 800 V! Portable equipment that uses tubes is almost unrealizable.

Figure 10.57 shows several types of tubes that can be found in old (and even modern) equipment. Some old types found in some military equipment use metal enclosures, replacing the glass, and some packages can contain multiple configurations (a triode and a pentode, for example).

Tubes are designated by a part number. Two main codes are used in their specfications. One code uses a number followed by a group of letters and then another number. The first number is the filament voltage, or the voltage that must be applied to the filament to heat the tube. The second and third sets of digits indicate the specific function. For instance, a 12AX7 is a double triode tube that has a 12 V filament.

The other code system is European, where a group of letters indicates the function of the tube, and the final number designates the specific type. For example, the ECC83 is a double triode (CC) tube (equivalent to the 12AX7!).

The best method for identifying a tube is to consult a manual; Prompt Publications has an excellent replacement guide for tubes (*Tube Substitution Handbook,* Delmar Learning, ISBN 0-7906-1036-1).

Tube Testing

When the filament opens (burns), the tube will no longer function. In most cases, a malfunction of any tube-based equipment is caused by a burned-out tube. One

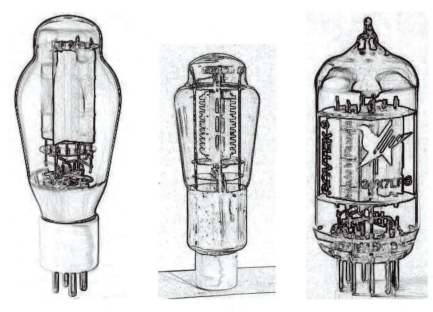

Figure 10.57

advantage of tubes from the technician's point of view is that you can see if a tube is burned out—when the equipment is turned on, you will not observe the weak orange or red light usually emitted by its filament. A visual inspection can reveal problems in this case. Unfortunately, this is not conclusive; in some equipment, the filaments are wired in series. In that case, if one of them burns out, the others receive no power and also do not function.

You can test the filament of a tube using a multimeter. A fully operational test requires special circuits or test equipment.

CHAPTER 11

Power Supplies

GENERAL CONSIDERATIONS

All mechatronic projects must be powered by an energy source. Many different sources can be used (e.g., hydraulic, pneumatic, and even such alternatives as wind and heat), but the most important is electricity.

Most mechatronic projects are powered from cells or from the AC power line. As secondary energy sources, we can consider solar cells, fuel cells, and even small dynamos or alternators powered from gasoline or other motors.

In this chapter, we will provide basic information about energy sources that can be used to power mechatronic projects, with special emphasis on electrical supplies. Pneumatic, hydraulic, and alternative sources will be discussed but in a more limited manner. After all, *mechatronics* comes from *mechanics + electronics*.

CHEMICAL CELLS

Chemical cells convert power released in chemical reactions into electrical power. They are the most popular source of power in robots, since they are compact, cheap (in some cases), and can provide enough energy for most design requirements. We call the basic unit of a chemical energy source a *cell*. When multiple cells are wired together, they form a *battery*.

CELLS AND BATTERIES

Figure 11.1 shows the schematic symbols that are used to represent cells and batteries (with and without the internal resistance). Cells and batteries can be divided into two main groups.

1. *Common.*   When the chemical energy in a common cell or battery is exhausted, it must be discarded.

Figure 11.1

2. *Rechargeable.*   When the unit has discharged, and no chemical reaction can take place, it can be recharged and used again. Common rechargeable batteries and cells can last through more than 1,000 charge and discharge cycles.

CELL/BATTERY TYPES

The attributes of common types of cells and batteries are described below.

Zinc

- They are not fully rechargeable.
- They can be rejuvenated.[*]
- They are the most common and cheapest type.
- They are sold in cell sizes AAA, AA, A, C, and D, or as batteries (9-V or other).
- They are recommended for low-power (low-current) applications (see Fig. 11.2).

Alkaline

- They are not rechargeable.
- They cannot be rejuvenated (they may explode).

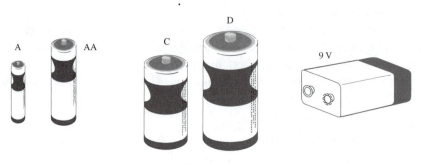

Figure 11.2

---

*Here, *rejuvenated* means that we can let the cell or battery "rest" for several hours to "recover its power," or we can wire it to a power supply to reactivate its internal reactive substances.

- They are sold in the same sizes as zinc cells.

- They are recommended for medium-power applications (including robotics and mechatronics).

- Their capacity increases when warmed and drops significantly at temperatures below freezing.

Nickel-Cadmium (Nicad)

- They are rechargeable up to 1,000 times.

- They are sold in the same sizes as zinc cells, and in other formats for special applications (see Fig. 11.3).

- They can provide up to three times the energy of an equivalent zinc cell.

- They are recommended for medium- and high-power applications.

- Special types can provide up to three times the energy of common Nicad cells.

- Nicad batteries have less overall capacity than lead-acid batteries of the same dimension (see Fig. 11.3).

- Nicad cells can be subject to a phenomenon called *memory effect* if only partially discharged before recharging. The term refers to the cells' apparent ability to "remember" the depth of the charge they received and accept only that much charge in the next cycle.

Lead-Acid

- They are rechargeable (the type found in motorcycles and cars).

- They are sold in 6- and 12-V versions.

- They are recommended for high-power applications.

- They are available in sealed versions.

- They are very heavy, increasing the weight of a robot or other mobile project.

Figure 11.3

Gel-Cell

- They are rechargeable.

- They are sold in special formats that are ideal for robotic and mechatronic applications—similar to lead-acid.

- They are more expensive than lead-acid.

- They are sold in 6- and 12-V versions (see Fig. 11.4).

When using a cell or battery, we should consider the following characteristics.

Electromotive Force (Voltage)

Electromotive force (EMF) refers to the voltage found across the cell or battery terminals. Although many types are specified with typical values of 1.5 V (zinc and alkaline cells), 1.2 V (Nicad), 9 V, or 12 V, the correct (measured) value can be higher in a fully charged unit. During discharge, the voltage provided by a cell or battery falls, as shown in Fig. 11.5. A battery or cell is considered to be "dead" when its voltage falls to 80% of the nominal voltage.

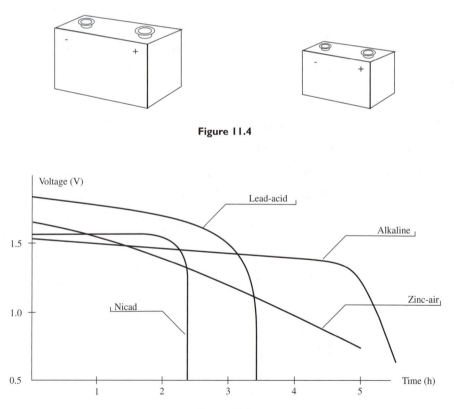

Figure 11.4

Figure 11.5

Amp-Hour Current

Amp-hour (Ah) current is the specification that refers to the amount of energy that a cell or battery can produce. For instance, a 12 V × 10 Ah battery can deliver

- 10 A current for 1 hr

- 5 A current for 2 hr

- 2 A for 5 hr

Formula:

The discharge cycle (in hours) of a battery or cell can be computed as

$$t = E/I \tag{11.1}$$

where t = discharge time (in hours)
          E = discharge rate of the battery (Ah)
          I = current delivered to the load (A)

Internal Resistance

All electrical energy sources have an internal resistance that "absorbs" part of the energy generated by the device. This internal resistance (r) in series with the resistance of the load (R) forms a voltage divider. Therefore, as shown in Fig. 11.6, as the current provided by the cell rises, the voltage across the load falls; the higher the internal resistance, less current the cell can deliver to a load without affecting the voltage across it. This means that high-power cells and batteries are likely to have low internal resistance.

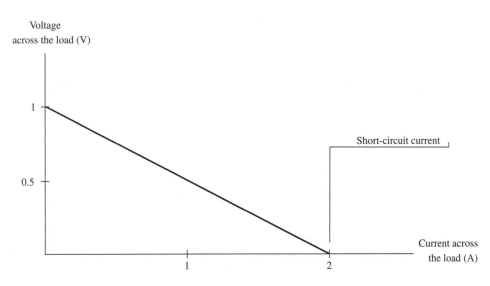

Figure 11.6

The internal resistance of a cell or battery normally is specified by the manufacturer. An indirect way to specify the internal resistance is by giving the short-circuit current.

Formulas:

- Current delivered to a load (circuit shown in Fig. 11.7),

$$I = V/(r + R) \tag{11.2}$$

- Voltage across a load (Fig. 11.8),

$$V_L = V[R/(R + r)] \tag{11.3}$$

- Short-circuit current (Fig. 11.9),

$$I = V/r \tag{11.4}$$

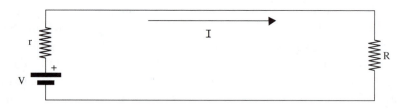

Figure 11.7

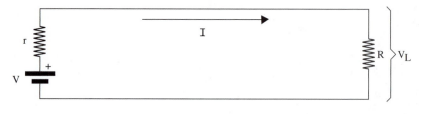

Figure 11.8

Figure 11.9

where V = voltage of the cell or battery (V)
            R = load resistance (Ω)
            r = internal resistance of the cell or battery (Ω)
            I = current in across the circuit (A)
          V_L = voltage across the load (V)

The amp-hour ratings of some common cells are given in Table 11.1. The characteristics of some other types of cells are provided in Table 11.2.

Table 11.1   Ah Ratings of Common Cells[*]

Battery type	Capacity (mAh)	Typical drain (mA)
AAA, zinc	400	5
AAA, alkaline	1,000	10
AA, zinc	800	20
AA, Alkaline	2,000	50
C, zinc	2,500	40
C, alkaline	6,000	100
D, zinc	5,000	80
D, alkaline	12,000	200

[*] Varies by manufacturer.

Table 11.2   Characteristics of Other Types of Cells[*]

Type	Voltage (V)	Power density	Internal resistance
Lithium, NR	3.0	very high	high
Nicad, R	1.2	moderate	low
Lead-acid, R	2.0	moderate	low
Nickel-hydride, R	1.2	high	–
Lithium polymer, R	–	high	–
Sodium sulfur, R	–	high	–
Zinc-air, R	1.5	very high	high
Zinc-bromine, R	–	high	–
Carbon-zinc, NR	1.5	low	high
Alkaline, NR	1.5	high	high

[*] NR = not rechargeable, R = rechargeable.

In Fig. 11.10, we show the recommended discharge curves of some batteries and cells as function of their size (capacity) and discharge current. This should be helpful in choosing the proper battery.

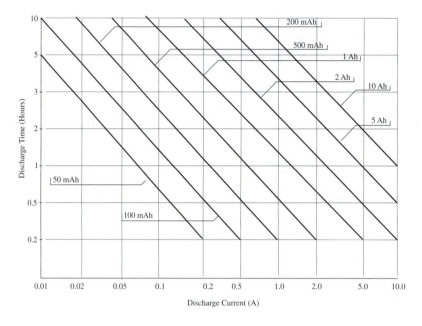

Figure 11.10

Recharge Rate

Most batteries must be recharged slowly, over a period of 10 to 26 hr. If too much current passes through the battery during the recharge process, the generated heat can damage it, or it may even explode! The recommended current limit for a recharge process is one-tenth of the ampere-hour rating of the cell or battery. This means that a safe recharge current for a 1000-Ah Nicad cell is 100 mA.

RECHARGING CIRCUITS

A rechargeable cell or battery is recharged by passing a current across it in the reverse of the normal flow direction to a load, as shown in Fig. 11.11. The simplest battery charger is a voltage source that uses a resistor as current limiter, as shown in Fig. 11.12. The resistor value is calculated using Eq. (11.5).

Figure 11.11

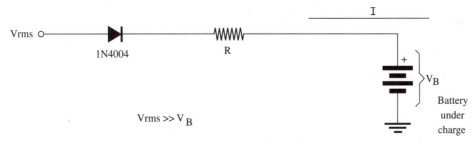

Figure 11.12

Formulas:

Value of R

$$R = \frac{V_{RMS}}{I} \qquad (11.5)$$

where R = resistor value in (Ω)
V = RMS voltage of the transformer's secondary coil (V)
I = charging current (A)

Dissipation of R

$$P = R \times I^2 \qquad (11.6)$$

where P = dissipation power (W)
R = resistor value in (Ω)
I = charging current (A)

The designer should use a resistor with at least twice the calculated dissipation power.

A more complex battery charger is a constant-current source such as shown in Fig. 8.5, in Chapter 8. Resistor value R is calculated to allow the charging current to flow across the battery under charge.

INTERCONNECTION OF CELLS

If we need to increase the power provided by cells or batteries, they can be connected in series or parallel. Parallel connection is recommended only if the cells or batteries have identical characteristics.

Series Connections

When cells or batteries are connected in series, the current capacity stays the same, but the voltage delivered to the load becomes the sum of the voltages of all of the interconnected units, as suggested in Fig. 11.13. The total internal resistance is the sum of the internal resistances of the associated batteries or cells.

Formulas:

a. Voltage of cells/batteries connected in series,

$$V = V_1 + V_2 + V_3 + \dots + V_n \qquad (11.7)$$

where V = total voltage (V)

$V_1,\dots,V_n$ = voltages of the associated cells or batteries (V)

b. Equivalent internal resistance,

$$r = r_1 + r_2 + r_3 + \dots\dots + r_n \qquad (11.8)$$

where r = equivalent internal resistance (Ω)

$r_1,\dots,r_n$ = internal resistances of the associated cells or batteries (Ω)

Parallel Connections

Parallel connection of cells or batteries increases their current capacity but maintains a constant voltage across the load. The internal resistance is reduced. Figure 11.14 shows how to connect batteries or cells in parallel.

Obs.: All of the cells or batteries must be of the same voltage.

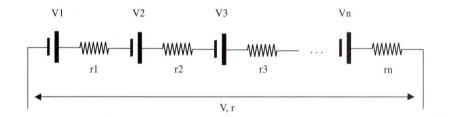

Figure 11.13

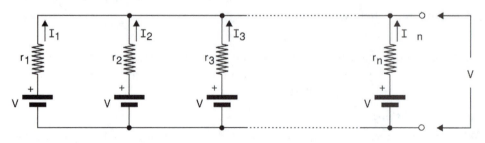

Figure 11.14

Formulas:

- Current capacity,

$$I = I_1 + I_2 + I_3 + ... + I_n \tag{11.9}$$

where I = total current capacity of the batteries (A)
$I_1,..., I_n$ = current capacities of the associated cells or batteries (A)

- Internal resistance,

$$1/r = 1/r_1 + 1/r_2 + 1/r_3 + ... + 1/r_n \tag{11.10}$$

where r = total internal resistance (Ω)
$r_1,...,r_n$ = internal resistances of the associated cells or batteries (Ω)

Series/Parallel Connections

Batteries and cells can be connected in a combination of series and parallel as shown in Fig. 11.15. In this configuration, both current and voltage capacities are increased.

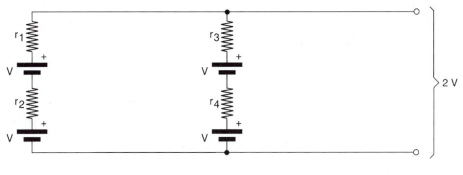

Figure 11.15

FUEL CELLS

Fuel cells convert the combustion energy of a substance (i.e., its reaction with oxygen) into electricity. Fuel cells are not common in robotic or mechatronic applications, because they are new, expensive, and primarily in an developmental phase. They are used to power experimental vehicles as cars and buses, and they have applications in space, where they provide energy to satellites and space probes.

The main types of fuel cells are as follows:

- *Alkaline cells.* These were used by NASA in early space missions, operating with an efficiency of about 70%. They are too expensive for mass applications.

- *Phosphoric acid fuel cells.*   These have a major disadvantage: phosphoric acid is dangerous.

- *Molten carbonate fuel cells.*   The main disadvantage is that these must operate in a 1,200°F (650°C) ambient.

- *Solid oxide fuel cell.*   These provide high power and efficiency of about 60%. They are not suitable for use in smaller applications.

- *Proton exchange membrane (PEM) fuel cells.*   PEMs are the most promising in terms of cost, efficiency, and producibility. One disadvantage is that they release hydrogen into the environment.

- *Metal-air fuel cells.*   In this type, the fuel is metal, which makes it environmentally friendly and highly versatile. Metal-air cells hold great promise for future applications.

SOLAR CELLS

An interesting source of electrical energy for mechatronic projects is solar power. The sun can provide about 1 kW/m^2 in some parts of the world. This energy is free, but, unfortunately, there is no efficient way to convert it to electricity. Solar cells are made with silicon, forming panels that typically have very low (less than 10%) efficiency. Since the panels are not cheap, they are not a good alternative for powering robots or other mechatronics devices unless other energy sources are unavailable.

Solar cells can be wired in series, parallel, or series/parallel, in the same manner as other electrical sources.

ALTERNATIVE ELECTRICAL ENERGY SOURCES

Figure 11.16 shows some alternative electric energy sources: (a) wind, (b) hydraulic, (c) gravity, and (d) spring power.

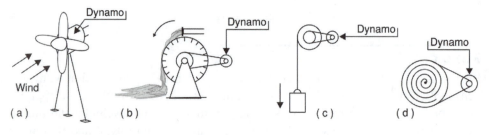

Figure 11.16

POWER SUPPLIES (AC POWER LINE)

The most important power source available in our home or lab is the AC power line. It is used to power home appliances, instruments, and electrically driven machinery in industrial plants. The AC power line provides an alternating voltage of 110, 117, 127, 220, or 240 V, depending on the country or region. Unfortunately, most of the small motors used to power robots and mechatronic devices are DC types that need lower voltages, and the electronic circuits normally operate on low-voltage DC.

To convert AC line power into DC, we use circuits called *power supplies*. The simplest power supplies have the configuration shown in Fig. 11.17. A voltage regulator or current regulator is optional.

HALF-WAVE RECTIFIERS

This rectifier (see Fig. 11.18) uses one diode and furnishes a DC pulsed voltage at the frequency of AC input. The following formulas are used to calculate the current and voltage across a load.

Formulas:

$$U_{avg} = \frac{U_{max}}{\pi} \qquad (11.11)$$

where U_{avg} = average voltage across the load in volts (V)
        U_{max} = voltage peak or maximum in volts (V)
        $\pi = 3.1416$

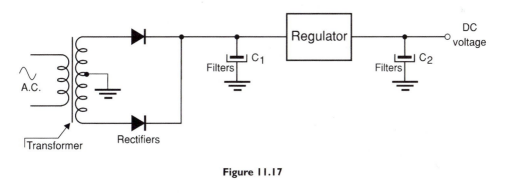

Figure 11.17

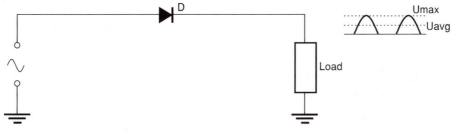

Figure 11.18

Obs.: The voltage drop in the diodes (about 0.2 V in germanium types and 0.6 V in silicon types) is not considered.

$$I_{avg} = \frac{I_{max}}{\pi}$$ (11.12)

where I_{avg} = average current through the load in amperes (A)
I_{max} = peak or maximum AC input current in amperes (A)
$\pi = 3.1416$

FULL-WAVE RECTIFIERS

The current and voltage in a full-wave rectifier can be calculated using the following formulas. These are valid for both two-diode systems and bridges (see Fig. 11.19).

Formulas:

$$U_{avg} = \frac{2 \times U_{max}}{\pi}$$ (11.13)

where U_{avg} = median voltage across the load in volts (V)
U_{max} = peak voltage or maximum voltage in volts (V)
$\pi = 3.1416$

$$I_{avg} = \frac{2 \times I_{max}}{\pi}$$ (11.14)

where I_{avg} = average current in the load measured in amperes (A)
I_{max} = peak current or maximum current in amperes (A)
$\pi = 3.1416$

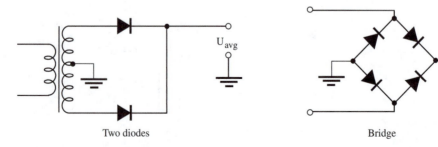

Two diodes   Bridge

Figure 11.19

Derived formula:

$$I_{max} = \frac{I_{avg} \times \pi}{2} \tag{11.15}$$

Obs.: The voltage drop in the diodes is not considered in this calculation.

LC FILTER COEFFICIENT

The filter coefficient is defined as the ratio of the amplitude of the alternating component at its input to the amplitude of the alternating component at the output. The filter coefficient is calculated by the following formulas (see Fig. 11.20).

Formulas:

$$\alpha = \frac{U_{win}}{U_{wout}} \tag{11.16}$$

where α = filter coefficient
U_{win} = alternating component at the filter's input in volts (V)
U_{wout} = alternating component at the filter's output in volts (V)

$$\alpha = \omega^2 CL - 1 \tag{11.17}$$

where α = filter coefficient
$\omega = (2 \times \pi \times f)$, where f is the frequency of the alternating current at the filter's input (Hz)
C = capacitance in farads (F)
L = inductance in henrys (H)

RC FILTER COEFFICIENT

The filter coefficient of an RC configuration is defined as the ratio of U_{wout} to U_{win} of the LC filter. The following formulas are used to calculate this coefficient (see Fig. 11.21).

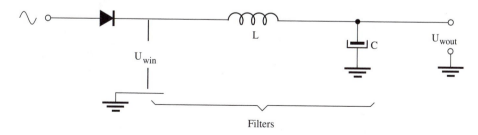

Filters

Figure 11.20

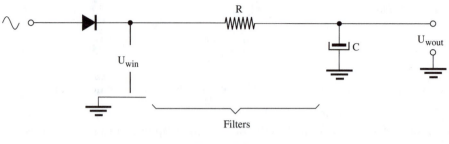

Figure 11.21

Formulas:

$$\alpha = \frac{U_{wout}}{U_{win}} \tag{11.18}$$

where α = filter coefficient
U_{wout} = amplitude of the alternating component at filter's output in volts (V)
U_{win} = amplitude of the alternating component at the filter's input in volts (V)

$$\alpha = \sqrt{(\omega \times R \times C)^2 + 1} \tag{11.19}$$

where α = filter coefficient
$\omega = 2 \times \pi \times f$, where f is the frequency at the filter's input in hertz (Hz)
R = resistance in ohms (Ω)
F = capacitance in farads (F)

RIPPLE FACTOR

The ripple factor (γ) is defined as the ratio of the rms value of the output voltage to the DC value of the output voltage times 100. The following formula expresses this relationship and is used to calculate the ripple factor (see Fig. 11.22):

Formula:

$$\gamma = \frac{U_{rms}}{U_{dc}} \tag{11.20}$$

where γ = ripple factor
U_{rms} = rms value of the voltage at the output in volts (V)
U_{dc} = average value of the output voltage in volts (V)

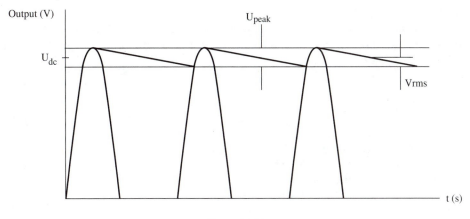

Figure 11.22

Ripple Factors (Resistive Load)

a. Half-wave rectifier = 120%

b. Full-wave rectifier = 48%

Table 11.3 shows various rectifier characteristics.

Table 11.3 Rectifier Characteristics (Using Resistive Loads)

	Half-wave	Full-wave (center tap transformer)	Full-wave (bridge)
U_{dc} in the load	$\dfrac{U_{max}}{\pi}$	$\dfrac{2 \times U_{max}}{\pi}$	$\dfrac{2 \times U_{max}}{\pi}$
U_{rms} in the load	$\dfrac{U_{max}}{2}$	$\dfrac{U_{max}}{\sqrt{2}}$	$\dfrac{U_{max}}{\sqrt{2}}$
Reverse voltage accross the diodes, $U_r = $ (max)	U_{max}	$2 \times U_{max}$	U_{max}
Ripple factor (γ)	120%	48%	48%
Transformer: storage factor referred to the output power in the load[*]	$3.49 \times P_{dc}$ (primary and secondary)	$1.75 \times P_{dc}$ (secondary) $1.23 \times P_{dc}$ (primary)	$1.23 \times P_{dc}$ (primary and secondary)

[*] P_{dc} = output power in watts (W).

FILTER INDUCTANCE

The following formula is used to calculate the inductance in a filter. This values depends on the filter coefficient, the capacitor employed, the output current, and the frequency applied to the filter (see Fig. 11.23).

Formula:

$$L = \frac{1 + \alpha}{\omega^2 \times C} \tag{11.21}$$

where L = inductance in henrys (H)
   α = filter coefficient
   $\omega = 2 \times \pi \times f$, where f is the frequency in hertz (Hz)
   C = capacitance in farads (F)

Obs.: We can define the quantity η as the filter efficiency, and the value is given by the following formula:

$$\eta = \frac{1}{\alpha} \tag{11.22}$$

where α = filter coefficient
   η = filter efficiency

FILTER CAPACITANCE

Filter capacitance can be calculated as a function of the filter inductance and its AC frequency and filter coefficient. The following formula can be used for this calculation:

Formula:

$$C = \frac{1 + \alpha}{\omega^2 \times L} \tag{11.23}$$

where C = filter capacitance in farads (F)
   α = filter coefficient
   $\omega = 2 \times \pi \times f$, where f is the AC frequency in hertz (Hz)
   L = coil inductance in henrys (H)

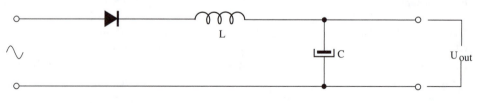

Figure 11.23

Derived Formula

$$C = \frac{1 + \eta}{\eta \times \omega^2 \times L} \tag{11.24}$$

where η = filter efficiency

All other variables are as defined in Eq. (11.23).

PRACTICAL CIRCUITS

Some practical circuits are shown in Fig. 11.24. The output voltage ranges (or fixed value) and the maximum currents are as listed below.

Figure 11.24a.　Simple power supply, 6 or 12 V @ 500 mA

Figure 11.24b.　Unregulated power supply, 0 to 12 V @ 1 A

Figure 11.24c.　Regulated power supply, 6 or 12 V @ 1 A

Figure 11.24d.　Variable power supply using the LM350T, 3 A

Figure 11.24e.　12 V @ 10 A power supply using the 2N3055 (heavy duty)

VOLTAGE MULTIPLIERS

The input AC voltage can be increased after rectification to produce a higher DC voltage in the output of the circuit. This can be accomplished by circuits called *voltage multipliers.* Several are described below.

Conventional Voltage Doubler

Figure 11.25 shows the basic configuration for a conventional voltage doubler. The working voltages of the capacitors are calculated as shown below.

Formula:

$$V_{C1} = V_{C2} = V_{max} \tag{11.25}$$

where　V_{C1} and V_{C2} = capacitor working voltage in volts (V)

Obs.: The minimum peak inverse voltage (PIV) of the diodes is $2 \times V_{max}$.

Cascade Voltage Doubler

This circuit (see Fig. 11.26) uses only two diodes. The capacitors' working voltages are calculated as shown below.

Formula:

$$V_{C1} = V_{max} \qquad V_{C2} = 2 \times V_{max} \tag{11.26}$$

where　V_{C1} and V_{C2} = capacitors' working voltages in volts (V)

V_{max} = peak or maximum value of the AC input voltage in volts (V)

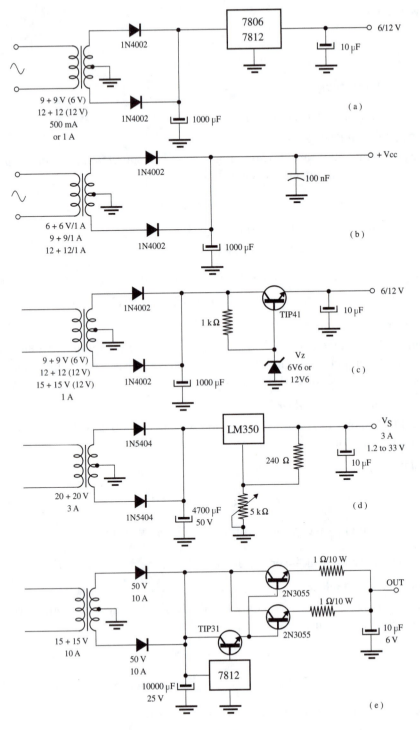

Figure 11.24

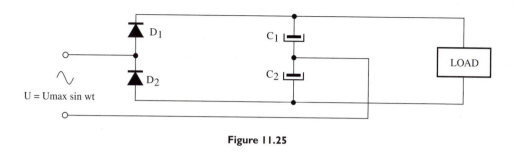

Figure 11.25

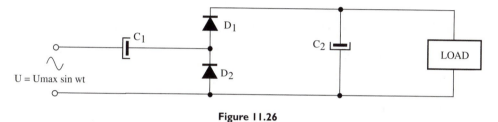

Figure 11.26

Obs.: The minimum PIV of the diodes is $2 \times V_{max}$.

Bridge Voltage Doubler

Figure 11.27 shows a configuration using four diodes. Here, the output voltage is twice the input voltage.

Formula:

$$V_{C1} = V_{C2} = V_{max} \qquad (11.27)$$

where V_{C1} and V_{C2} = minimum DC working voltage of the capacitors in volts (V)

V_{max} = peak value of the applied input voltage in volts (V)

Obs.: The minimum PIV of the diodes is $2 \times V_{max}$.

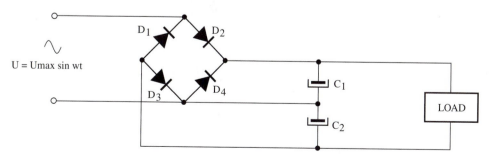

Figure 11.27

Full-Wave Tripler

The circuit shown in Fig. 11.28 produces an output that is three times the input voltage. Component values are calculated as shown below.

Formula:

$$V_{C1} = V_{C3} = V_{max}$$

$$V_{C2} = 2 \times V_{max}$$

(11.28)

where V_{c1}, V_{c2}, V_{c3} = minimum DC working voltages of the capacitors in volts (V)

V_{max} = peak AC input voltage in volts (V)

Obs.: The minimum PIV of the diodes is $2 \times V_{max}$.

Cascade Voltage Tripler

The circuit shown in Fig. 11.29 has a voltage output that is three times the input voltage. The capacitors' characteristics are calculated as shown below.

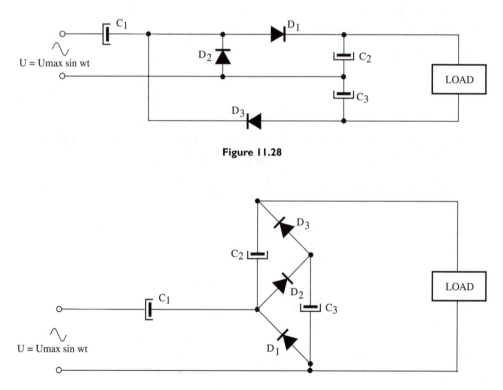

Figure 11.28

Figure 11.29

Formula:

$$V_{C1} = V_{max}$$

$$V_{C2} = V_{C3} = 2 \times V_{max}$$

(11.29)

where $V_{c1}...V_{c3}$ = minimum DC working voltages of the capacitors in volts (V)

V_{max} = peak AC input voltage in volts (V)

Obs.: The minimum PIV of the diodes is $2 \times V_{max}$.

Full-Wave Voltage Quadrupler

Figure 11.30 shows a circuit that uses four diodes. The minimum DC working voltages of the capacitors can be calculated as shown below.

Formula:

$$V_{C1} = V_{C2} = V_{C3} = V_{C4} = 2 \times V_{max}$$

(11.30)

where $V_{c1}... V_{c4}$ = minimum DC working voltages of the capacitors in volts (V)

V_{max} = peak AC input voltage in volts (V)

Obs.: The minimum PIV of the diodes is $2 \times V_{max}$.

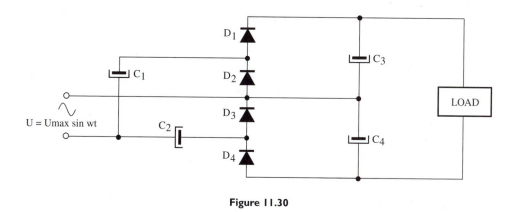

Figure 11.30

VOLTAGE REGULATORS

The purpose of a voltage regulator is to maintain a constant voltage in a circuit. Many equipment power supplies use this type of IC to ensure a regulated output supply voltage.

Today, many appliances use a three-terminal, 1-A voltage regulator such as the 78XX and 79XX series. [In this series, the XX is replaced by a number that indicates the output voltage. For example, the 7806 provides a 6-V output under 1 A of current (maximum).]

Another type of voltage regulator can be found in switched mode power supplies (SMPSs) such as used in computers, video monitors, robots, and many other modern appliances.

The operating principles of these regulators differ from those of linear regulators, as they acts as oscillators, producing a voltage that depends on the frequency and pulse width of the output signal.

Specifications

Voltage regulator ICs, as in the case of any other ICs, are specified by part number. The electrical characteristics are found in data provided by the manufacturer. The main specifications of ICs of this type are outlined below.

  a. *Input voltage range.*   This indicates the range of voltages that can be applied to the input. In the documentation, it is often abbreviated as V_{in}.

  b. *Output voltage or range (V_{out}).*   Some ICs are fixed-voltage regulators. They have an internal zener that sets the output voltage. Others are controlled by an external zener or resistors to establish the desired output voltage.

  c. *Output current.*   It is important to observe the maximum output current that can be regulated by the IC.

Table 11.4 provides performance data for some common voltage regulator ICs (see also Fig. 11.31).

Table 11.4   Common Voltage Regulators

Type	Input voltage range (V)	Output voltage (V)	Maximum output current (A)
7806—positive	8–18	6	1
7808—positive	10–18	8	1
7812—positive	14–18	12	1
7906—negative	8–18	6	1
7908—negative	10–18	8	1
7912—negative	14–18	12	1
LM150/250350-T—variable	4–35	1.25–33 V	3

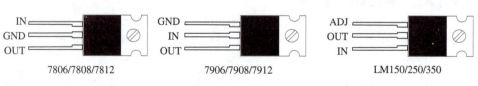

7806/7808/7812   7906/7908/7912   LM150/250/350

Figure 11.31

POWER BOOSTERS

Many devices (e.g., DC motors and solenoids), when powered up, momentarily draw a current that can be many times that of the normal operating current (see Fig. 11.32). If a power supply with a high internal resistance (or low current capacity) is the power source for these devices, the voltage across them when they are switched on can drop to a value that makes start-up impossible. Motors can stall, and solenoids can fail to operate.

There are several ways to boost the power supply output under such conditions, allowing it to provide a high momentary current. The most common way is to install a large-value capacitor in parallel with the power supply, as shown in Fig. 11.33.

Another way to boost a DC motor or solenoid is by momentarily adding a second voltage source, or paralleling a second source, as shown in Fig. 11.34. The switch used to activate the booster supply can be replaced by a relay's contacts.

Formula:
The energy stored in a capacitor can be calculated as follows:

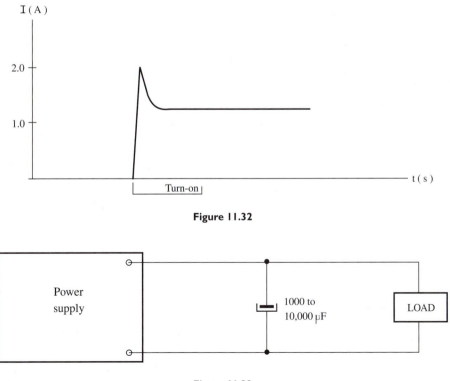

Figure 11.32

Figure 11.33

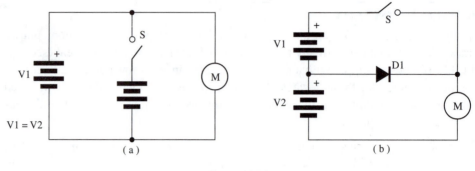

Figure 11.34

$$E = \frac{1}{2} \times C \times V^2 \qquad (11.31)$$

where E = stored energy in joules (J)
C = capacitance of the capacitor in farads (F)
V = voltage between the plates of the capacitor in volts (V)

HYPERCAPACITORS

New technologies are resulting in very high-capacitance capacitors built into small packages. Capacitors in the farad range are now available, and they can be used to boost power supply outputs to heavy-duty loads for several seconds.

A hypercapacitor can deliver a current of more than 100 A for a few seconds, which is sufficient to start a stalled, powerful DC motor. This eliminates the need for heavy-duty batteries. Figure 11.35 shows how a hypercapacitor can be used to start a motor or to drive a very high-current load for several seconds.

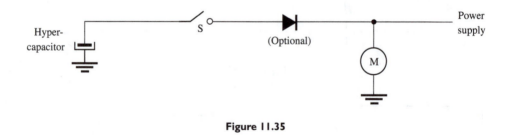

Figure 11.35

PNEUMATIC SOURCES

Air (or any other gas) under pressure can provide power. A small compressor can be used to deliver air under pressure to move small motors or pistons as shown in Fig. 11.36. A polyethylene terephthalate (PET) bottle can be used as a reservoir of pressurized air to move air muscles or pistons in the arrangement shown in Fig. 11.37.

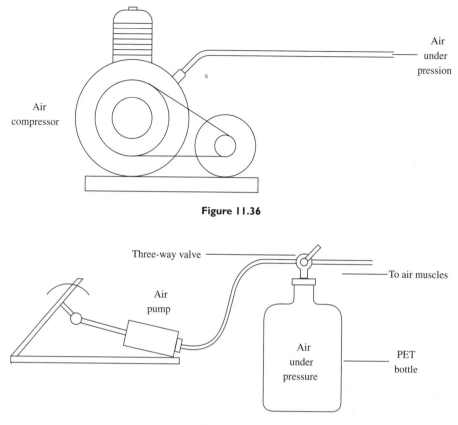

Figure 11.36

Figure 11.37

The maximum pressure of the stored air is reached when bottle deformation begins. Because the power depends on the pressure, and the pressure falls as air is delivered to the load, the typical power curve of an air reservoir used as pneumatic power source looks like the one shown in Fig. 11.38.

GASOLINE ENGINES

Internal combustion engines can provide power for many types of projects. The motors designed for model airplanes are suitable for many applications. The basic question is how to couple them to the mechanism, since they run at very high rpm. Controlling their speed is very difficult.

THERMAL ENERGY SOURCES

Small steam engines can be a practical power source for experimental mechatronic projects. Figure 11.39 shows how a small steam engine can be built from a few parts.

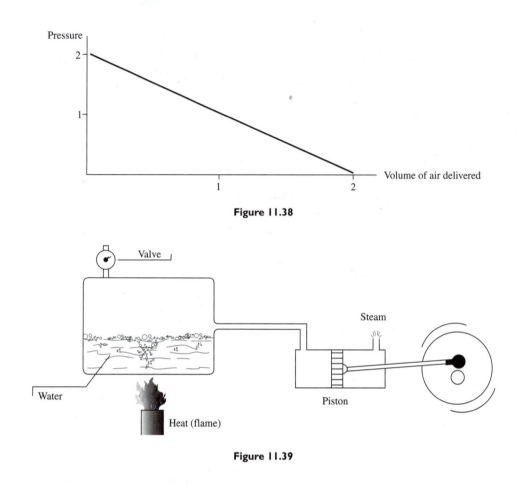

Figure 11.38

Figure 11.39

SPRINGS AND RUBBER BANDS

A contracted or expanded spring is a potentially useful energy source. If coupled to a suitable mechanism, it can release enough energy to produce motion. The amount of energy stored in a spring can be calculated as shown in Chapter 2. The same principle is valid for a twisted or expanded rubber band, as shown in Fig. 11.40.

Of course, the energy stored in a spring or rubber must be drawn from an external source, which reinforces the concept that such devices can store energy but cannot create it. Keep this in mind if you consider using them as power sources.

Twisted rubber band

Figure 11.40

CHAPTER 12

Microcontrollers, Microprocessors, and Digital Signal Processors

---

ADDING INTELLIGENCE

Two forms of intelligence can be added to mechatronic projects: rule-based and neural. In some projects, the two forms can work together.

THE BASIC STAMP

The BASIC Stamp product line is a family of small computers that run Parallax BASIC (PBASIC) programs. They can be used for direct control of TTL logic and other devices or to acquire data from sensors or control units via their I/O ports.

Light emitting diodes, pushbuttons, reed switches, sensors, and driver stages such as described in Chapters 4–6 and 8–10 can be connected directly to the I/O ports. The BASIC Stamp can also be connected to RS-232 networks, allowing it to share data with remote units using a two-wire connection.

A BASIC Stamp can be used as the "brain" in robots, mechatronic devices, and other mechanisms, where it controls motion, makes decisions (according to its programming or based on information picked up by sensors), and interacts with a human operator.

ARCHITECTURE

Parallax Inc. offers the BASIC Stamp in several physical versions, but all have the same logical design as shown in Fig. 12.1. They consist of the following basic blocks:

- A 5-V regulator
- A resonator

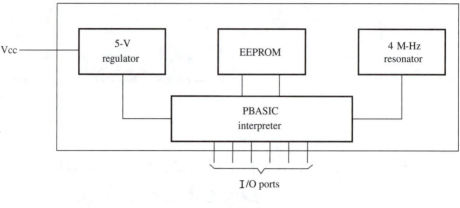

Figure 12.1

- A serial electrically erasable programmable read-only memory (EEPROM)
- A PBASIC interpreter

The interpreter chip fetches the instructions one at a time and performs the appropriate operations on the I/O pins or internal structures within the interpreter. The EEPROM stores the PBASIC program, and the chip can be programmed and reprogrammed a virtually infinite number of times without removing it from the computer to erase it.

PROGRAMMING

To program the BASIC Stamp, you just have to connect it to an IBM PC or compatible and run the software to edit and download the program.

The BASIC Stamp 1 is connected to the PC via serial port, and the BASIC Stamp 2 and 2sx via parallel port. Figure 12.2 shows how to connect the BASIC Stamp to a PC for programming.

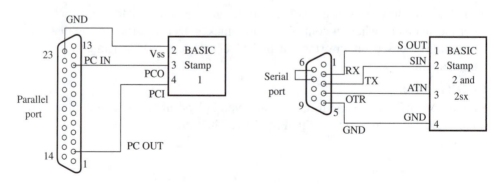

Figure 12.2

VERSIONS

The BASIC Stamp can be obtained in several versions, three of which are shown in Fig. 12.3. The features of each are outlined below.

BASIC Stamp 1

- Provides 8 I/O pins
- Capacity of 80 to 100 instructions
- Includes 256-byte EEPROM
- Provides seven word registers; all byte addressable, two bit addressable
- Execution rate of 2,000 instructions/s

This version is ideal for projects with tight space limitations.

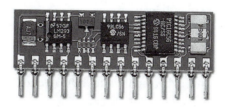

BS1

BS2

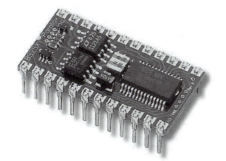

BS2sx

Figure 12.3 *(Courtesy of Parallax, Inc.)*

BASIC Stamp 2

- Provides 16 I/O pins plus 2 synchronous serial pins
- Capacity of 500 to 600 instructions
- Includes 2048-byte EEPROM
- Provides 16 word registers; all byte, nibble, and bit addressable
- Execution rate of 4,000 instructions/s

This is the most popular version, used in hobby, educational, and industrial applications; suitable for both DOS and Windows programming.

BASIC Stamp 2sx

- Provides 16 I/O pins plus 2 synchronous serial pins
- Capacity of 500 to 600 instructions
- Includes 16 kB EEPROM (program and data in 8 × 2k program spaces)
- Execution rate of 10,000 instructions/s
- Provides 16 word registers; all byte, nibble, and bit addressable, plus 64 bytes (1 for program ID and 63 for user)

PACKAGE TYPES

The BASIC Stamp is available in four package types.

1. PC board (original version with a prototype area and battery clips)
2. and 3. OEM BASIC Stamps 1 and 2 (recommended for an OEM design using Parallax interpreter chips)
4. Surface mount (14- or 24-pins versions; these provide the smallest possible package)

The general characteristics of these three styles are shown in Table 12.1

APPLICATIONS

Figure 12.4 shows some circuits driven by the outputs of a BASIC Stamp and some sensors connected to the inputs. More examples can be found in the previously suggested chapters.

Additional application suggestions, data sheets, and programming assistance can be found at http://www.parallaxinc.com/.

PROGRAMMABLE INTELLIGENT CONTROLLERS

Some useful programmable intelligent controllers (PICs) are manufactured by Microchip Technology Inc. (http://www.microchip.com). The common features of

Table 12.1 BASIC Stamp Characteristics

Feature	BASIC Stamp 1	BASIC Stamp 2	BASIC Stamp 2sx
Program execution speed	4 MHz	20 MHz	50 MHz
Current requirements	2 mA running, 20 μA sleep mode	7 mA running, 50 μA sleep mode	60 mA running, 200 μA sleep mode
Serial I/O speed	TTL serial @ 300–2400 baud (I/O pins)	TTL and RS-232 serial @ 200–50k baud (I/O pins or serial port) (0–19,200 baud with flow control)	TTL and RS-232 serial @ 305–115k baud (I/O pins or serial port) (0–19,200 baud with flow control)
Processor	PIC16C56	PIC16C57	SX8AC
Programming interface (PC)	Parallel port	Serial port	Serial port
I/O instructions	BUTTON DEBUG HIGH INPUT LOW OUTPUT POT PULSIN PULSOUT PWM REVERSE SERIN SEROUT SOUND TOGGLE	BUTTON COUNT DEBUG DTMFOUT FREQOUT HIGH INPUT LOW OUTPUT PULSIN POULSOUT PWM RCTIME REVERSE SERIN SEROUT SHIFTIN TOGGLE XOUT	BUTTON COUNT DEBUG DTMFOUT FREQOUT HIGH INPUT LOW OUTPUT PULSIN POULSOUT PWM RCTIME REVERSE SERIN SEROUT SHIFTIN TOGGLE XOUT

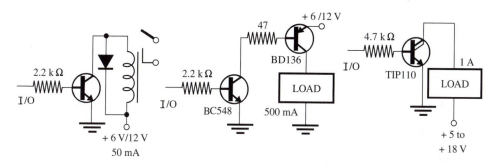

Figure 12.4

these microcontrollers are that all are available in one-time programmable (OTP) packages, and they use the same instruction set. All the members of the family use the same architecture (Harvard).

Obs.: The term *architecture* refers the way the microcontroller or computer solves problems. There are many types, including RISC, Harvard, Von Neumann, and others. Some common ones are as follows:

- *RISC* stands for *reduced instruction set computer.* This indicates that the set of instruction is small, containing only the most useful ones. RISC instructions are typically executed in a single clock tick.

- *CISC* refers to a *complex instruction set computer* and is an architecture that includes more powerful instructions—but many of them need several clock cycles to execute.

- The Harvard architecture uses a scheme in which the program memory and data memory are not shared and use separate buses. This allows the program to be fetched and executed simultaneously with the data. PICs use a RISC architecture.

PICs can be used to add intelligence to mechatronic projects, including robots. You can use them to process data picked up by sensors or sent by a keyboard, and they will produce control signals for motors, solenoids, and other devices.

SELECTING A PIC

There are more than 130 different types of PICs, offering a variety of memory configurations, low voltage and power, and small footprints. All of them are very easy to use.

Working from the specifications of a project, it is easy to find a PIC to match. In its web site, Microchip offers a parametric search engine to help the user find the correct PIC for a particular application.

The basic criteria to be used in the selection process are as follows:

- Program memory size
- Data RAM size
- Memory type
- Package

Microchip offers five families of eight-bit MCUs with the following characteristics:

PIC12CXXX

- RISC architecture
- 8-pin package (DIP and SOIC)

- 12-bit/14-bit program word
- 2.5 V operating voltage
- EEPROM data memory
- Interrupt handling

PIC16C5X

- 12-bit program word
- 14-, 18-, 20-, and 28-pin packages (SOIC and SSSOP)
- Low-voltage operation down to 2.0 V
- Ideal for battery-operated applications
- Operating voltage up to 15 V, allowing use with batteries

PIC16CXXX

- 14-bit program word
- Analog-to-digital converter at 12 bits
- 18- to 64-pin packages
- 14-bit instruction set
- 8-level hardware stack

PIC17CXXX

- 16-bit program word
- RISC architecture
- 16-bit instruction word
- Powerful vectored interrupt-handling capabilities

PIC18CXXX

- Enhanced 16-bit program word
- CMOS
- Integrated analog-to-digital converter (ADC)
- RISC architecture
- 32-level-deep stack
- Multiple interrupt sources
- 77 instructions (total)
- Performance of 10 MIPS

- Programmable low-voltage detect (LVD)

- Programmable brownout detect (BOD)

Figure 12.5 shows a popular PIC and its pin functions.

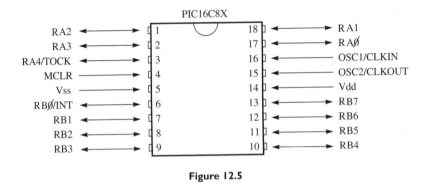

Figure 12.5

INSTRUCTION SET

There are only 35 instruction in the RISC architecture of the PIC chips. More complex instructions are built on these simple ones. Many macros are included in the Microchip standard assembler to help the designer create a richer instruction set (see Table 12.2). More information about PIC chips can be found at http://www.microchip.com.

Obs.: One of the most popular series of PIC microprocessors today is the 16F8X.

PROGRAMMING AND USING PICS

Like other microcontrollers, PICs are programmed using a PC or Macintosh computer. A small board with a socket is connected to the parallel or serial port, and emulator software is used to create the program and transfer it to the chip as shown in Fig. 12.6. The software to create the program can be downloaded from the web site of the manufacturer.

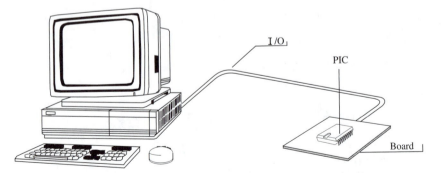

Figure 12.6

Table 12.2 PIC Instruction Set

Mnemonic, operands	Description	Mnemonic, operands	Description
ADDWAF f,d	Add W and f	IORWF f,d	Inclusive OR E with f
ANDWF f,d	Add W with f	MOVF f,d	Move f
CLRF f	Clear f	MOVWF f	Move W to f
CLRW	Clear W	NOP	No operation
COMF f,d	Compliment	RLF f,d	Rotate left f trough carry
DECF f,d	Decrement	RRF f,d	Rotate right through carry
DECFSZ f,d	Decrement, skip if 0	SUBWF f,d	Subtract W from f
INCF f,d	Increment f	SWAPF f,d	Swap nibbles in f
INCFSZ f,d	Increment, skip if 0	XORWF f,d	Exclusive OR W with f
Bit-oriented file register operations			
BCF f,b	Bit Clear f	BTFSC f,b	Bit test f, skip if clear
BSF f,b	Bit set f	BTFSS f,b	Bit test f, skip if set
Literal and control operations			
ADDLW k	Add literal and W	RETFIE	Return from interrupt
ANDLW k	AND literal with W	RETLW k	Return with literal in W
CALL k	Call subroutine	RETURN	Return from subroutine
CLRWDT	Clear Watchdog Timer	SLEEP	Go into standby mode
GOTO k	Go to address	SUBLW k	Subtract W from literal
IORLW k	Inclusive OR literal with W	XORLW k	Exclusive OR literal with W
MOVLW k	Move literal to W		

Once programmed, the chip is installed in the application, where it controls motors, solenoids, and other devices based on signals it picks up from sensors. Since the device is TTL compatible, is can be used with many of the circuits suggested in this book to drive motors, solenoids, relays, and so on.

THE 80C51

One of the most popular of all microcontroller families is the 80C51. Introduced by Philips about 15 years ago, the 80C51 is an 8-bit microcontroller with 4k/8k OPT/ROM. The main characteristics of this microcontroller are as follows:

- Low-voltage operation, 2.7 to 5.5 V @ 16 MHz
- $4k \times 8$ ROM
- 128×8 RAM
- Three 16-bit counter/timers
- Boolean processor

- Memory addressing capability of 64 ROM and 64 RAM
- CMOS and TTL compatibility
- Two speed ranges at Vcc = 5 V: 0 to 16 or 0 to 33 MHz
- Security bits: ROM (2 bits) and OTP/EPROM (3 bits)

Figure 12.7 shows a block diagram of the 80C51. Figure 12.8 shows the three packages available for the 80C51 chips from Philips Semiconductor as follows:

1. Dual in-line package (DIP)
2. Plastic leaded chip carrier (PLCC)
3. Plastic quad flat pack (PQFP)

THE 8051 FAMILY

Table 12.3 reveals various characteristics of the 8051 family.

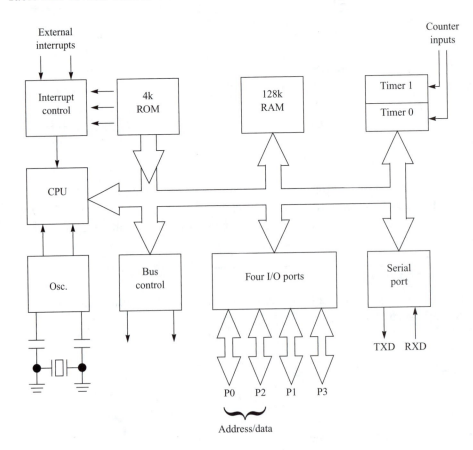

Figure 12.7

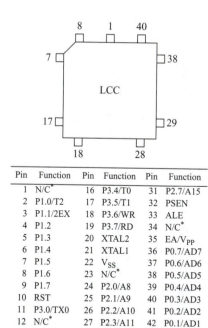

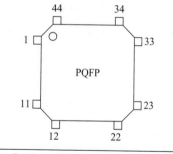

Pin	Function	Pin	Function	Pin	Function
1	N/C*	16	P3.4/T0	31	P2.7/A15
2	P1.0/T2	17	P3.5/T1	32	PSEN
3	P1.1/2EX	18	P3.6/WR	33	ALE
4	P1.2	19	P3.7/RD	34	N/C*
5	P1.3	20	XTAL2	35	EA/V$_{PP}$
6	P1.4	21	XTAL1	36	P0.7/AD7
7	P1.5	22	V$_{SS}$	37	P0.6/AD6
8	P1.6	23	N/C*	38	P0.5/AD5
9	P1.7	24	P2.0/A8	39	P0.4/AD4
10	RST	25	P2.1/A9	40	P0.3/AD3
11	P3.0/TX0	26	P2.2/A10	41	P0.2/AD2
12	N/C*	27	P2.3/A11	42	P0.1/AD1
13	P3.1/TX0	28	P2.4/A12	43	P0.1/AD1
14	P3.2/INT0	29	P2.5/A13	44	V$_{CC}$
15	P3.3/NT1	30	P2.8/A14		

* No internal connection

Pin	Function	Pin	Function	Pin	Function
1	P1.5	16	V$_{SS}$	31	PO.6/AD6
2	P1.6	17	N/C*	32	P0.5/AD5
3	P1.7	18	P2.0/A8	33	P0.4/AD4
4	RS7	19	P2.1/A9	34	P0.3/AD3
5	P3.0/RXD	20	P2.2/A10	35	P0.2/AD2
6	N/C*	21	P2.3/A11	36	P0.1/AD1
7	P3.1/TXD	22	P2.4/A12	37	P0.0/ADO
8	P3.2/INTO	23	P2.5/A13	38	V$_{CC}$
9	P3.3/NT1	24	P2.6/A14	39	N/C*
10	P3..4/TD	25	P2.7/A15	40	P1.0/T2
11	P3.5/T1	26	PSEN	41	P1.1/T2EX
12	P3.6/WR	27	ALE	42	P1.2
13	P3.7/RD	28	N/C*	43	P1.3
14	XTAL2	29	EA/VPP	44	P1.4
15	XTAL1	30	P0.7/AD7		

* No internal connection

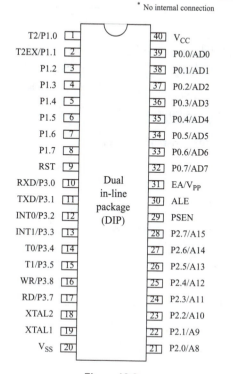

Figure 12.8

Table 12.3   8051 Characteristics

Device	ROMless version	EPROM version	ROM bytes	RAM bytes	16-bit timers
8051	8031	–	4k	128	2
80C51	80C31	87C51	44k	128	2
8052	8032	–	8k	256	3
80C52	80C32	87C52	8k	256	3
83C053	–	87C054	8k	192	2
83CL410	80CL410	–	4k	128	2
83C451	80C451	87C451	4k	128	2
83C528	80C528	87C528	32k	512	3 + WD*
83C550	80C550	87C550	4k	128	2 + WD
83C552	80C552	87C552	8k	256	3 + WD
83C562	80C562	–	8k	256	3 + WD
83C652	80C652	87C652	8k	256	2
83C654	–	87C654	16k	256	2
83C751	–	87C751	2k	64	1
83C752	–	87C752	2k	64	1
83C851	80C851	–	4k	128	2

* WD = watchdog.

Obs.: In addition to Philips, many other manufacturers have versions of the 8051 chip that include more memory and other features. By the time this book goes to press, Philips is certain to have new types that are not specified in the table.

The 8051 instruction set is as enumerated in Table 12.4.

Table 12.4   8051 Instruction Set

Mnemonic	Operation
Arithmetic instructions	
ADD A, <byte>	A = A + <byte>
ADDC A, <byte>	A = A + <byte> + C
SUBB A, <byte>	A = A – <byte> – C
INC A	A = A + 1
INC <byte>	<byte> = <byte> +1
INC DPTR	DPTR = DPTR + 1
DEC A	A = A – 1
DEC <byte>	<byte> = <byte> – 1
MUL AB	B:A = B x A
DIV AB	A = int (A/B)

Table 12.4 8051 Instruction Set (continued)

Mnemonic	Operation
	B = Mod (.B)
DA A	Decimal adjust

Logical instructions

ANL A, <byte>	A = A.AND. <byte>
ANL <byte>, A	<byte> = <byte>. AND. A
ANL <byte>, #data	<byte> = <byte> AND.#data
ORL A, <byte>	A = A.OR.<byte>
ORL <byte>, #data	<byte> = <byte>. OR. #data
ORL <byte>, A	<byte> = <byte> . OR.A
XRL A, <byte>	A = A.XOR. <byte>
XRL <byte>, #data	<byte> = <byte>. XOR.#data
XRL <byte>, A	<byte> = <byte>. XOR. A
CRL A	A = 00H
CPL A	A = .NOT.A
RL A	Rotate ACC left 1 bit
RLC A	Rotate left through carry
RR A	Rotate ACC right 1 bit
RRC A	Rotate right through carry
SWAP A	Swap nibble in A

Data transfer instruction that access internal memory space

MOV A, <src>	A =<src>
MOV <dest>, A	<dest> = A
MOV <dest>, <src>	<dest> = <src>
MOV DPTR, #data16	DPTR = 16-bit immediat constant
PUSH <src>	INC SP:MOV"@SP",<src>
POP <dest>	MOV <dest>, "@SP".DEC SP
XCH A, <byte>	ACC and <byte> exchange data
XCHD A, @Ri	ACC and @Ri exchange low nibbles

Lookup table read instructions

MOV C, @A+DPTR	Read program memory at (A + DPTR)
MOV C, @A+ PC	Read program memory at (A + PC)

Boolean instructions

ANL C.Bit	C = C.AND.bit
ANL C,/bit	C = C.AND.NOT bit
ORL C.bit	C = C.OR.bit
ORL C,/bit	C = C.OR.NOT bit
MOV C,bit	C = bit

Table 12.4   8051 Instruction Set (continued)

Mnemonic	Operation
MOV bit, C	bit = C
CLR C	C = 0
CLR bit	bit = 0
SETB C	C = 1
SETB bit	bit = 1
CPL C	C = NOT.C
CPL bit	bit = NOT.bit
JC rel	Jump if C = 1
JNC rel	Jump if C = 0
JB bit,rel	Jump if bit = 1
JNB bit,rel	Jump if bit = 0
JBC bit,rel	Jump if bit = 1, CLR bit

Unconditional jumps

Mnemonic	Operation
JMP addr	Jump to addr
JMP @A, DPTR	Jump to A + DPTR
CALL addr	Call subroutine at addr
RET	Return from subroutine
RET1	Return from interrupt
NOP	No operation

Conditional jumps

Mnemonic	Operation
JZ rel	Jump if A = 0
JNZ rel	Jump if A≠0
DJNZ <byte>,rel	Decrement and jump if not zero
DJNE A, <byte>, rel	Jump if A≠<byte>
CJNE <byte>, #data, rel	Jump if <byte>≠#data

COP8

The COP8 is a line of microcontrollers from National Semiconductor. It is divided into two families called the *Basic Family* and the *Feature Family*.

The Basic Family members are intended for use in lower-end, lower-cost applications requiring less memory and using simpler peripheral devices. Conversely, the Feature Family members are designed for applications that require more memory and use more advanced peripheral devices. Both have the same basic architecture and basic instruction set, and they can share common peripherals. More information can be found at http://www.national.com.

The COP8 microcontroller can be found in the following versions:

a. COP8 FLASH—virtual EEPROM, USART, dual clock, and 2.5–5.5 V ISP (no external voltages):

- COP8SBR9—32k flash, 1 k RAM, 2.7–2.9 V brownout
- COP8SBE9 —8k version COP8SBR9 with 256 RAM
- COP8SCR9—32k Flash, 1k RAM, 4.17–4.5 V brownout
- COP8SCE9 —8k version of COP8SCR9 with 256 RAM
- COP8SDR9—32k flash, 1k RAM, no brownout
- COP8SDE9—8k version of COP8SDR9 with 256 RAM
- COP8CBR9—COP8SBR9 plus 10-bit A/D, 2.7–2.9 V brownout
- COP8CBE9—8k version with 256 RAM
- COP8CCR9—COP8SCR9 plus 10-bit A/D, 4.17–4.5 V brownout
- COP8CCE9—8k version of COP8CCR9 with 256 RAM
- COP8CDR9—COP8SDR9 plus 10-bit A/D, no brownout
- COP8CDE9—8k version of COP8CDR9 with 256 RAM
- COP8CFE9—COP8CDE9 without USART or dual clock
- COP8AME9—8k flash, 512 RAM, 10-bit A/D, 2 op amps, temp. sensor, 4.7–4.5 V brownout
- COP8ANE9—COP8AME9 with no brownout

b. COP8 OPT—low EMI; zero external components; 2.7 V–5.5 V; 28/44-pin CSP packages—factory programmed ROM versions

- COP8SAA7—1k OTP, 1 timer, power on reset
- COP8SAA5—ROM version of COP8SAA7
- COP8SAB7—2k OTP, 1 timer, power on reset
- COP8SACB5—ROM version of COP8SAB7
- COP8SAC7—4k OTP, 1 timer, power on reset
- COP8SAC5—ROM version of COP8SAC7
- COP8SGE7—8k OTP, 2 comparators, USART, 3 timers
- COP8SGE5—ROM version of COP8SGE7
- COP8SGR7—32k OTP, 2 comparators, USART, 3 timers
- COP8SGR5—ROM version of COP8SGR7
- COP8ACC7—16k OTP, 12-bit A/D, comparator, volt. ref, 2.5–5.5 V
- COP8ACC5—ROM version of COP8ACC7

c. ROM—2.5–5.5 V

- COP912C—768 ROM, 64 RAM, 1 timer
- COP8LXXCJ RJ—4k OTP version
- COP820C/840C—1k or 2k ROM, 64/128 RAM, 1 timer
- COP87LXXCJ RJ—4k OTP version
- COP820CJ—1k ROM, 64 RAM, comparator, brownout
- COP87LXXCJ RJ—4k OTP version

- COP840CJ—2k ROM, 128 RAM, comparator, brownout
- COP8LXXCJ RJ—4k OTP version
- COP880C—4k ROM, 128 RAM, 1 timer
- COP8SEC5—4k ROM, 128 Bytes EERAM
- COP8SEC7—16k OTP version
- COP884BC—4k ROM, 2.0B CAN, 1 timer, 12 interrupters
- COP87L84BC—16k OTP version
- COP888EB—8k ROM, 2.0B CAN, 8-bit A/D, SPI, USART
- COP87L88RB—32k OTP version
- COP888CF—4k ROM, 8-bit A/D, 2 timers, 10 interrupts
- COP97L88CF—16k OTP version
- COP888CL—4k ROM, 2 timers, 10 interrupts
- COP87L88CL—16k OTP version

Obs.: Other types are listed in National's web site, along with additional information.

PROGRAMMING AND USING THE COP8

COP8 devices are available in 28-pin CSP, 44-pin LLP, and 48- and 56-pin TSSOP packages. To develop a new project using one of the devices, you will need an interface board to connect to your computer. This board will allow you to design the project using emulator software running on your computer.

National Semiconductor offers many hardware emulators. For simple projects, the recommended ones are the COP8Flash Start Kit and the COP8 OTP/ROM Prototyping Kit. Figure 12.9 shows a typical application of a COP8 in mechatronics.

THE MSP430

The MSP430 family of ultra-low-power 16-bit RISC mixed-signal processors from Texas Instruments provides an excellent solution for battery-powered applications. These processors include digital and analog signal processing. The primary features are as follows:

- Ultra-low power architecture:
    - 0.1 µA RAM retention
    - 0.8 µA real-time clock mode
    - 250 µA/ MIPS active
- High-performance analog, ideal for precise measurements

Figure 12.10 shows a typical application for this microcontroller.

OTHER MICROCONTROLLERS AND MICROPROCESSORS

Many other devices can be used to add intelligence to our projects. These include the following:

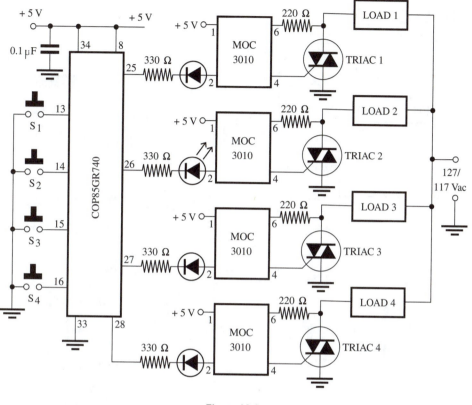

Figure 12.9

- The Z80.   This old device, being one of the simplest and easiest to use, is still used today for teaching microprocessor applications.

- Old mother boards.   Old 286, 386, and 486 mother boards can be used to add simple controls to mechatronic devices.

Although these computers no longer useful for most general-purpose computing applications, they are ideal for experiments involving control, since they contain all the elements need to run the required software.

DIGITAL SIGNAL PROCESSORS

Most of our mechatronic projects work with signals provided by analog sensors. This means that, when we consider adding intelligence to these projects through the use of a microprocessor, the first problem to solve relates to signal compatibility.

Microprocessors and microcontrollers are digital devices, but sensors deliver analog signals. At the other end, once the signals have been processed by the intelligent circuits (i.e., the microprocessor or microcontroller), they are output in digital

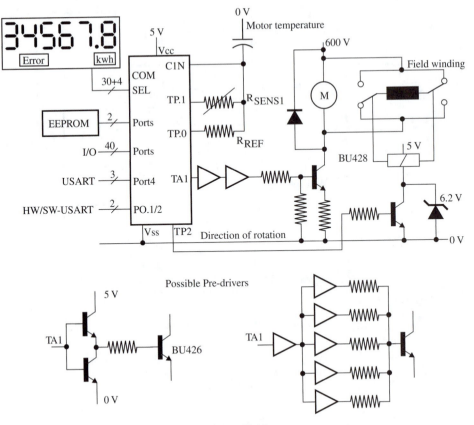

Figure 12.10

form. But most output devices require analog signals. We therefore need to be able to change analog signals to digital, and vice versa, using analog-to-digital (A/D) and digital-to-analog (D/A) converters.

A simple project that employs the structure shown in Fig. 12.11 can become very complex (needing many chips) if the traditional solution is given: A/D and D/A converters and a microprocessor. A better solution to this problem can be achieved with the use of a digital signal processor (DSP).

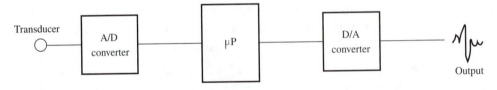

Figure 12.11

Digital signal processors are devices that process analog signals after converting them to digital form. After processing a signal, they can convert it back to an analog output as shown in Fig. 12.12.

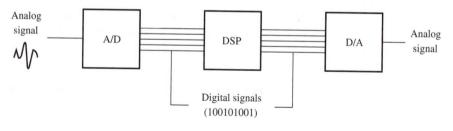

Figure 12.12

DSPs differ from microprocessors in a number of ways. Microprocessors are designed for a range of general-purpose functions, running large blocks of software. The DSP is a single-minded specialist running a small range of functions at lightning speeds. They are used as embedded controllers, which are devices that contain all the necessary software for a particular group of tasks.

Designed to be fast, DSPs can work with signals from most types of sensors (sound, light, temperature, pressure, images, speed, and so forth), giving quick instructions to control motors, solenoids, lights, heaters, and other devices.

Many families of DSPs can be used in mechatronic projects. One of the most popular is the TMS320 from Texas Instruments (TI).

The DSPs from TI can be programmed using a simple emulator plugged into the appropriate port of a PC. Their I/O capabilities allow them to be connected directly to sensors and devices that are being controlled.

USING A DSP

DSPs are designed for different applications, depending on their dynamic range. This range is specified as the spectrum of numbers, from small to large, that the DSP has to process in an application. Any DSP device must have the capacity to work with whatever numbers are applied to its input.

The basic applications for DSPs in mechatronic projects are as follows:

- Motor controls (AC, DC, and brushless)
- Braking systems
- Inverters
- Sensor interfaces
- Pumps and fans
- Servo motion control

DSPs are used in the same manner as microcontrollers: you need a programming interface and software to transfer the program, as appropriate to the application. The emulator hardware is connected to the PC or Mac via I/O ports (serial or parallel), and the DSP is plugged into a socket.

Most DSP manufacturers sell their emulators in many versions, including a starter kit for beginners. The software can be downloaded from manufacturers' web sites or obtained on CD.

THE TMS320 FAMILY

Texas Instruments has three families of DSPs, each with distinctive instruction set architectures.

1. TMS320C6000.   Highest performance; these are the fastest DSPs, with scalable performance up to 1.1 GHz.

2. TMS320C5000.   Best power efficiency; they offer low power consumption— as low as 0.09 V and 0.05 mW/MIPS. Performance is up to 800 MIPS. They are ideal for portable products.

3. TMS320C2000.   Control optimized; these provide a unique combination of processing power and control-specific peripherals. They are used in applications employing motors and industrial automation (mechatronics), HVAC, and in uninterruptible power supplies (UPS).

Detailed information about the DSPs from TI, including tutorials and technical documentation, can be found at http://www.ti.com.

OTHER DSPS

Many manufacturers have DSPs in their products line. Examples include:

- Lucent Technologies, DSP32C

- Analog Devices, ADSP2181, 21060, and so on

- Motorola, DSP56002, DSP56156, DSP96002, and others

CHAPTER 13

Interfacing with a Computer

CONTROLLING DEVICES

A personal computer and its I/O ports provide an inexpensive and very powerful platform for implementing projects involving control of real-world mechatronic devices. The standard PC has two types of ports that can be used for control and data acquisition: parallel and serial.

Both types have their advantages and disadvantages, which will be clear when the reader compares the characteristics and applications of each as described in this chapter. Therefore, the information contained herein will be useful when designing any mechatronic project that is controlled or interfaced with a PC. Of course, the same concepts are valid for other types of computers such as the Macintosh.

THE PARALLEL PORT

The parallel port (also known as the line printer terminal, LPT) provides eight TTL outputs, five inputs, and four bidirectional channels.

The advantages are

- It is easy to use.
- It allows the data to be transmitted simultaneously on all data and control lines.
- It is very fast.

The disadvantages are

- It requires a multiwire line.
- The transmission line must be very short.

THE PARALLEL PORT PINOUT

Access to the parallel ports of a PC is made by a DB-25 male connector as shown in Fig. 13.1. The pinout scheme is shown in Table 13.1.

Table 13.1   Parallel Port Pin Configuration

Pin	Description	PC function	Pin	Description	PC function
1	Strobe	Output	10	ACK	Input
2	Data 0	Output	11	Busy	Input
3	Data 1	Output	12	Paper empty	Input
4	Data 2	Output	13	Select	Input
5	Data 3	Output	14	Auto feed	Output
6	Data 4	Output	15	Error	Input
7	Data 5	Output	16	Initialize printer	Output
8	Data 6	Output	17	Select input	Output
9	Data 7	Output	18–25	Ground	–

The printer cable uses the DB-25 connector at one end, but it has a Centronics connector (female) at the other, as shown in Fig. 13.2. The pin arrangement is shown in Table 13.2.

OUTPUT VOLTAGES AND CURRENTS

The outputs of the parallel port are TTL compatible. This means that they output logic in the following form:

Output logic level	Voltage
1	5 V
0	GND (0 V)

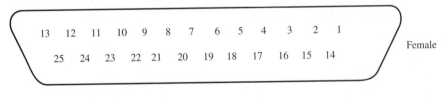

Figure 13.1

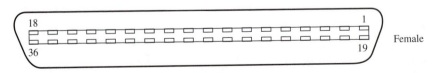

Figure 13.2

Table 13.2 Centronics Connector (Female) Pin Configuration

Pin	Function	Pin	Function	Pin	Function
1	Strobe	10	ACK	18	–
2	Data 0	11	Busy	19–29	Ground
3	Data 1	12	Paper out	30	–
4	Data 2	13	Select out	31	Printer reset
5	Data 3	14	Autofeed	32	Error
6	Data 4	15	–	33	Ground
7	Data 5	16	–	34	–
8	Data 6	17	Ground	35	Select in
9	Data 7				

The ports can drive TTL inputs directly, but the amount of current drained or delivered through an output depends on the load.

Figure 13.3 is a graph of the output voltage plotted against the current provided by the output. It is easy to see that if the impedance is below 2.2 kΩ, the voltage across the port will not be high enough to be sensed as a "HI" logic level. HI logic levels must be between 2.7 and 5 V, and LO logic levels must be between 0 and 0.8 V.

Note in the graph that the maximum current drained by each output is 24 mA, and the maximum current delivered by each output is about 2 mA. This means that, to drive any external load, the designer has two basic design alternatives:

1. *Driving the TTL inputs directly.* Many TTL devices can be connected directly to the outputs of the parallel ports as shown in Fig.13.4.

2. *Using drive stages.* For the drive stages, we have three options, depending on degree to which the load is to be isolated from the computer. It is clear that,

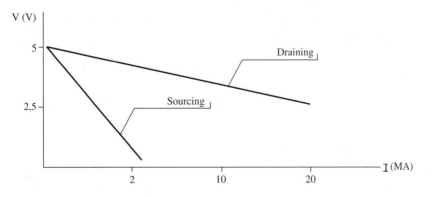

Figure 13.3

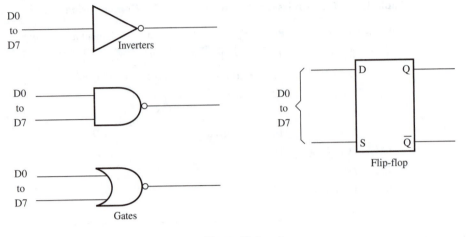

Figure 13.4

if the computer and the load (controlled circuit) are not electrically isolated, a fault in the load can damage the computer. The three isolation grades are defined in terms of the interfaces:

– *Isolation grade 0—directly coupled to the port.*   In this case, the controlled circuits are not isolated from the computer. In the case of any fault within the controlled circuit, the mother board of the computer, and many of its components, can be damaged. This kind of interface is recommended only for experienced users.

– *Isolation grade 1—using relays.*   In this type of interface, the controlled circuit is completely isolated from the computer, but the circuit used to drive the relay is not isolated from the computer. If something goes wrong with this intermediate stage, the mother board and associated components still can be damaged. This is recommended for the user who has some previous experience with this kind of circuit.

– *Isolation grade 2—using optocouplers.*   This is the safest solution. An optocoupler consists of an LED placed in front of a photo-sensor (photo-diode, photo-transistor, photo-diac, photo-trigger, etc.), with both sealed inside a hermetic package. When the LED is turned on, the control signal passes to the output device as a light beam. This provides total electrical isolation of the input signal and the output, as shown in Fig. 13.5. Typical opto-couplers can isolate voltages as high as 5,000 to 7,000 V.

PRACTICAL CIRCUITS

In the following text, we describe circuits that are suitable for interfacing a PC with mechatronic projects via the parallel port. The circuits are separated into two groups: Group 1 (isolation grades 0 and 1) and Group 2 (isolation grade 2).

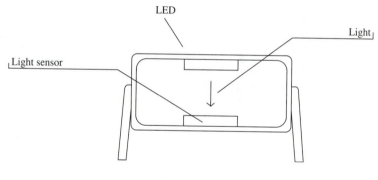

Figure 13.5

Since the PC parallel port can draw more current than it can deliver, it is possible to have two activation modes for the load: activated with the HI output level (noninverting) and activated with the LO output level (inverting).

Group I—Isolation Grades 0 and I

The circuits shown in Fig. 13.5 are considered to be isolation grade 0 when the load is connected directly to the control device (transistor, SCR, or other device) and isolation grade 1 when the load is replaced by a relay.

The circuits in Fig. 13.6 are as follows:

Figure 13.6a.　Low current (up to 2 mA) direct, noninverting

Figure 13.6b.　Low current (up to 20 mA) direct, inverting

Figure 13.6c.　Low power (up to 50 mA), one transistor, noninverting

Figure 13.6d.　Low power (up to 50 mA), one transistor, inverting

Figure 13.6e.　Medium power (up to 500 mA), two transistors, noninverting

Figure 13.6f.　Medium power (up to 500 mA), two transistors, inverting

Figure 13.6g.　High power (up to 3 A), Darlington transistor, noninverting

Figure 13.6h.　High power (up to 3 A), Darlington transistor, inverting

Figure 13.6i.　High power AC (up to 3 A), SCR, noninverting

Figure 13.6j.　High power AC (up to 3 A), SCR and transistor, noninverting

Figure 13.6k.　Low power (up to 10 mA), comparator or operational amplifier, inverting or noninverting

Figure 13.6l.　TTL buffer, inverting and noninverting

Figure 13.6m.　High-power IC, using TTL buffer

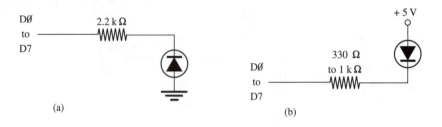

(a)

(b)

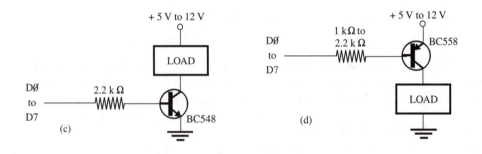

(c)

(d)

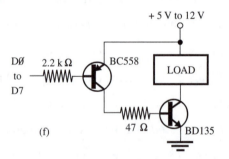

(e)

(f)

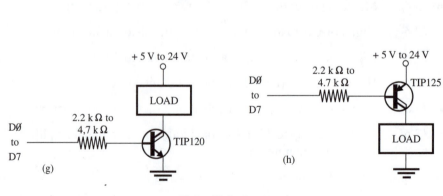

(g)

(h)

Figure 13.6　*(continues)*

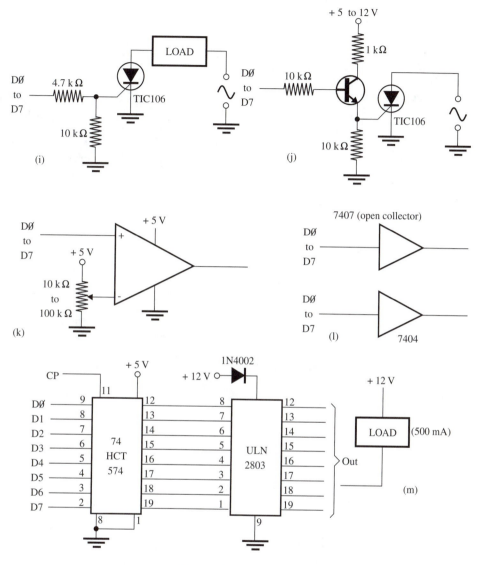

Figure 13.6 *(continued)*

Group 2—Isolation Grade 2

All of the circuits in Fig. 13.7 use optoisolators such as the 4N25 (or equivalent) or MOC3020 (for triacs), and all are fully isolated from the PC. The circuits can be associated with drive blocks as shown in Group 1 and Chapters 4 through 6.

Figure 13.7a. Direct

Figure 13.7b. Using a transistor

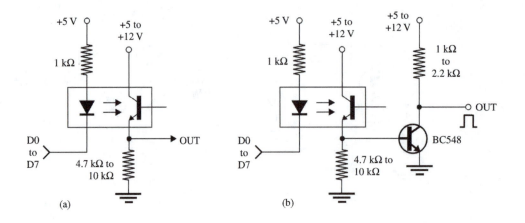

(a)

(b)

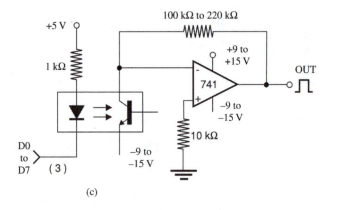

(c)

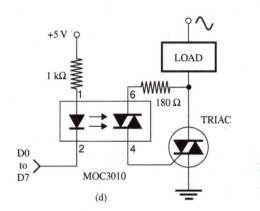

(d)

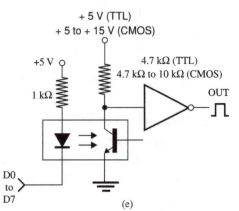

(e)

Figure 13.7

Figure 13.7c. Using an operational amplifier

Figure 13.7d. Triggering a triac

Figure 13.7e. Triggering logic

Types of Drive Configurations and Use

The proper use of the basic configurations described above depends on factors listed in Table 13.3.

Table 13.3 Drive Configurations and Applications

Circuit	Application
1. Inverting buffer	Driving loads requiring a few milliamperes. CMOS buffers suitable for this configuration: 4011, 4001, 4049.
2. Noninverting buffer	Driving loads requiring a few milliamperes without logic level inversion. CMOS buffers suitable for this application: 4050 or two 4011, 4001, or 4049 in cascade.
3. Optocoupled buffer	Used where isolation is need (isolation grade 2).
4. Medium-power transistor buffer	Used to drive loads up to 50 mA (depends on the gain of the transistor); inverting and noninverting configurations are possible. BC548, 2N2222 (NPN), and BC558 (PNP) are suitable for this task.
5. High-power transistor buffer	Used to drive loads up to several amps (depends on the transistor used). Inverting and noninverting configurations are possible. Two stages or Darlington transistors are used.
6. High-power circuit using an SCR	Nonisolated, high-power circuits working with AC or DC loads. Special attention to isolation is needed if circuit is powered from the AC line.

SOFTWARE

To command external devices that are connected to the parallel port, it is necessary to set the appropriate outputs to the HIGH or LOW logic level. This means that you need software to access the port. Some instructions are outlined below.

a. Output to address 278/378/3BC hex

This instruction transfers data from the data bus to the respective pins.

Bit	Pin
0	2
1	3
2	4
3	5
4	6
5	7
6	8
7	9

b. Output to address 27A/37A/3BE hex

This instructs the latch to capture the least significant bits of the data bus.

Bit	Pin
7	–
6	–
5	–
4	IRQ EN
3	17
2	16
1	14
0	1

These pins are driven by open-collector transistors pulled to +5 V through a 4.7 kΩ resistor. Each pin can sink about 7 mA and maintain 0.8 V at the low level.

c. Input from address 279/379/3BD hex

This command delivers real-time status to the processor from the pins according to the following table:

Bit	Pin
7	11
6	10
5	12
4	13
3	15
2	–
1	–
0	–

d. Input from address 27A/37A/3BE hex

This instruction makes the processor read the existing data on pins 1,14, 16, and 17, and the IRQ bit.

Bit	Pin
7	–
6	–
5	–
4	IRQ EN
3	17
2	16
1	14
0	1

QBASIC I/O PORT ACCESS

The INP function and OUT statements are used to access the I/O port as follows:

INF (portd)   'returns byte read from the port portd

OUT portd   'write byte value to the port portd

Obs.: (a) **Portd** can be any unsigned integer between 0 and 65535. (b) **Value** must be in the range of 0 to 255.

pdata = &H378

status = &H379

control = &H37A

OUT pdata, bits   'output data

bits = INF (status)   'input data

TURBO PASCAL I/O ACCESS

The predefined arrays Port and PortW are used to access the I/O ports as follows:

var Port: array [0 to 65535] of byte;

PortW: array [0 to 65534] of word

The value assigned to the PortW arrays is written out to the corresponding port. When the element of Port or PortW arrays is indicated by an expression, the value is read from the corresponding port.

Const Data =$378;

   Status = Data + 1;

   Control = Data + 2:

Var Bits: Byte;

Port (Data) : = Bits; (output data)

Bits ; (= Port(status)); (input data)

DEBUG I/O ACCESS

Running the debug:

C: Debug /?

Debug {[drive:] [path] filename [testfile-parameters]}

where,

[drive:] [path] filename—specifies the file to be tested

testfile-parameters—specifies the command line information needed by the file to be tested

Input and output commands:

i port_address

o port_address byte_value

C:\>debug

−o 378 5a

−i 379

7F

−q

C:\>

THE SERIAL PORT

The RS232 data interface provides the simplest way to establish two-way communications between a computer and a device or another computer.

RS-232 data is bipolar:

+3 to +12 V indicates ON or 0 state (space)

−3 to −12 V indicates OFF or 1 state (mark)

Many receivers are sensitive to differentials of 1 V or less.

Drivers supported by serial ports are as follows:

- Drivers requiring symmetrical power supplies (+ and − voltages) such as the 14488 series of ICs (many PCs use this IC)

- Low-power drivers requiring a 5-V power supply (they have an internal boost converter to rise the voltages)

- Low-voltage and low-power (3.3-V) devices meeting the EIA-562 standard (found in laptops and notebooks)

A typical UART[*] used as I/O chip for an RS-232 can deliver 50 mA per pin.

Pertinent RS-232 specifications are given in Table 13.4.

The serial port pin configuration (25-pin connector, see Fig. 13.8) is given in Table 13.5.

---

*Universal asynchronous receiver/transmitter. A UART is a microchip with programming to control the interface of a computer to attached serial devices.

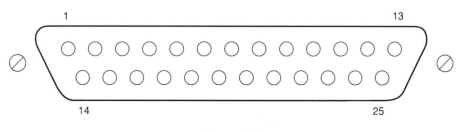

Figure 13.8

Table 13.4   RS-232 Specifications

Mode of operation	Single-ended
Total number of drivers and receiver on one line	1 driver, 1 receiver
Maximum cable length	50 ft
Maximum data rate	20 kb/s
Maximum driver output voltage	±25 V
Driver output signal level (loaded min)	±5 V to ±15 V
Driver output signal level (unloaded max)	±25 V
Driver load impedance	3 to 7 kΩ
Max driver current in high-Z state	±6mA @ ±2V
Slew rate (max)	30 V/ms
Receiver input voltage range	±15 V
Receiver input sensivity	±3 V
Receiver input resistance	3 to 7 kΩ

The serial port pin configuration (nine-pin connector, see Fig. 13.9) is given in Table 13.6.

Table 13.7 shows the arrangement of an AT Modem Standard cable with 9 to 25 pin connectors.

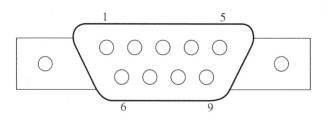

Figure 13.9

Table 13.5   Serial Port Pins, 25-Pin Connector

Pin	Function	
1	NC*	–
2	Transmitted data	TxD
3	Received data	RxD
4	Request to send	RTS
5	Clear to send	CTS
6	Data set ready	DSR
7	Ground	–
8	Carrier detect	CD
9–19	NC	–
20	Data terminal ready	DTR
21	NC	–
22	Ring indicator	RI
23–25	NC	–

* NC = not connected.

Table 13.6   Serial Port Pins, 9-Pin Connector

Pin	Function	
1	Carrier detect	CD
2	Received data	RxT
3	Transmitted data	TxD
4	Data terminal ready	DTR
5	Ground	–
6	Data set ready	DSR
7	Request to send	RTS
8	Clear to send	CTS
9	Ring indicator	RI

Table 13.7   AT Modem Standard Cable, 9 to 25 Pin Connectors

Description	Signal	9-pin DTE	25-pin DCE	Source, DTE or DEC
Carrier detect	CD	1	8	from modem
Receive data	RD	2	3	from modem
Transmit data	TD	3	2	from terminal/computer
Data terminal ready	DTR	4	20	from terminal/computer
Signal ground	SG	5	7	from modem
Data set ready	DSR	6	6	from modem
Request to send	RTS	7	4	from terminal/computer
Clear to send	CTS	8	5	from modem
Ring indicator	RI	9	22	from modem

PRACTICAL CIRCUITS

Power from the Computer

The interface circuits suggested herein can be powered from external power supplies (see Chapter 11) or draw power from the PC. Table 13.8 shows some points and the available voltage and current. See Fig. 13.10 for the pinout configuration.

Figure13.11 shows a typical RS-232 interface circuit using a UART chip.

Table 13.8 Connectors/Ports and Available Power

Location	Pins/connector	Power
Parallel port (Centronics)	Pin 18 = +5 V	100 mA (max)
Keyboard port	(a) 5-pin DIN pin 5 = +5 V pin 4 = GND (b) 6-pin mini DIN pin 1 = +5 V pin 2 = GND	200 mA
PS/S mouse port	Pin 1 = +5 V Pin 2 = GND	100 mA
Joystick port	Pins 1 and 9 = +5 V* Pins 4 and 5 = GND	Directly to the PC power supply—up to 20 A!
Disk drive connector	Pin 1 = +5 V Pin 4 = +12 V Pins 2 and 3 = GND	Directly to the PC power supply
Motherboard power connector	Pins 2, 10, 11, and 12 = +5 V Pin 3 = +12 V Pin 4 = –12 V Pin 9 = –5 V Pins 5 through 8 = GND	+ 5 V, up to 10 A +12 V, a few amps –5 and –12 V, max 100 mA (tip)
ISA bus	B3, B29 = +5 V B5 = –5 V B7 = –12 V B9 = + 12 V B1, B10 = GND	Same as the mother board connector

* Some PCI sound cards can supply 3.3 V instead of 5 V at the joystick connector; verify before use.

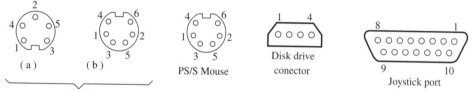

Figure 13.10

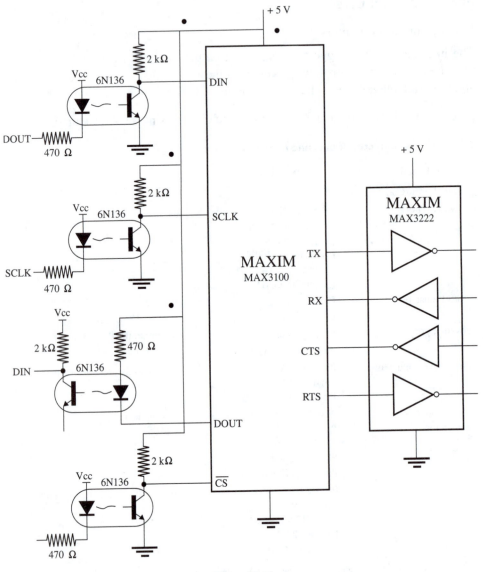

Figure 13.11

PCI Bus

The PCI local bus provides a high-performance connection to peripherals, expansion boards, and control boards (see Table 13.9). Voltage levels are 3.3 or 5 V.

Table 13.9 PCI Local Bus Configuration

Pin	5-V system environment		3.3-V system environment	
	Side B	Side A	Side B	Side A
1	–12 V	TRST#	–12 V	TRST#
2	TCK	+12 V	TCK	+12 V
3	GND	TMS	GND	TMS
4	TDO	TDI	TDO	TDI
5	+5 V	+5 V	+5 V	+5 V
6	+ 5 V	INTA#	+5 V	INTA#
7	INTB#	INTC#	INTB#	INTC#
8	INTD#	+5 V	INTD#	+5 V
9	PRSNT1#	Reserved	PRSNT1#	Reserved
10	Reserved	+5V (I/O)	Reserved	+3.3 V (I/O)
11	PRSNT2#	Reserved	PRSNT2#	Reserved
12	GND	GND	Con. Key[*]	Con. Key[*]
13	GND	GND	Con.. Key[*]	Con. Key[*]
14	Reserved	Reserved	Reserved	Reserved
15	GND	RST#	GND	RST#
16	CLK	+5 V (I/O)	CLK	+3.3 V (I/O)
17	GND	GNT#	GND	GNT#
18	REQ#	GND	REQ#	GND
19	+5 V (I/O	Reserved	+3.3 V (I/O)	Reserved
20	AD[31]	AD[30]	AD[31]	AD[30]
21	AD[29]	+3.3 V	AD[29]	+3.3 V
22	GND	AD[28]	GND	AD[28]
23	AD[27]	AD[26]	AD[27]	AD[26]
24	AD[25]	GND	AD[25]	GND
25	+3.3 V	AD[24]	+3.3 V	AD[24]
26	C/BE[3]#	IDSEL	C/BE[3]#	IDSEL
27	AD[23]	+3.3 V	AD[23]	+3.3 V
28	GND	AD[22]	GND	AD[22]
29	AD[21]	AD[20]	AD[21]	AD[20]
30	AD[19]	GND	AD[19]	GND
31	+3.3 V	AD[18]	+3.3 V	AD[18]
32	AD[17]	AD[16]	AD[17]	AD[16]
33	C/BE[2]#	+3.3 V	C/BE[2]#	+3.3 V
34	GND	FRAME#	GND	FRAME
35	IRDY#	GND	IRDY#	GND

Table 13.9　PCI Local Bus Configuration (continued)

Pin	5-V system environment		3.3-V system environment	
	Side B	Side A	Side B	Side A
36	+3.3 V	TRDY#	+3.3 V	TRDY#
37	DEVSEL#	GND	DEVSEL#	GND
38	GND	STOP#	GND	STOP#
39	LOCK$	+3.3 V	LOCK#	+3.3 V
40	PERR#	SDONE	PERR#	SDONE
41	+3.3 V	SBO#	+3.3 V	SBO#
42	SERR#	GND	SERR#	GND
43	+3.3 V	PAR	+3.3 V	PAR
44	C/BE[1]#	AD[15]	C/BE[1]#	AD[15]
45	AD[14]	+3.3 V	AD[14]	+3.3 V
46	GND	AD[13]	GND	AD[13]
47	AD[12]	AD[11]	AD[12]	AD[11]
48	AD[10]	GND	AD[10]	GND
49	GND	AD[09]	M66EN	AD[09]
50	C. Key[†]	C. Key[†]	GND	GND
51	C. Key[†]	C.. Key[†]	GND	GND
52	AD[08]	C/BE[0]#	AD[08]	C/BE[08]#
53	AD[07]	+3.3 V	AD[07]	+3.3 V
54	+3.3 V	AD[06]	+3.3 V	AD[06]
55	AD[05]	AD[04]	AD[05]	AD[04]
56	AD[03]	GND	AD[03]	GND
57	GND	AD[02]	GND	AD[02]
58	AD[01]	AD[00]	AD[01]	AD[00]
59	+5.0 V (I/O)	+5.0 V (I/O)	+3.3 V (I/O)	+3.3 V (I/O)
60	ACK64#	REQ64#	ACK64#	REQ64#
61	+ 5 V	+ 5 V	+ 5 V	+ 5 V
62	+ 5 V	+ 5 V	+ 5 V	+ 5 V
	C.Key[‡]	C.Key[‡]	C.Key[‡]	C.Key[‡]
	C.Key[‡]	C.Key[‡]	C.Key[‡]	C.Key[‡]
63	Reserved	GND	Reserved	GND[§]
64	GND	C/BE[7]#	GND	C/BE[7]#
65	C/BE[6]#	C/BE[5]#	C/BE[6]#	C/BE[5]#
66	C/BE[4]#	+5V (I/O)	C/BE[4]#	+3.3V (I/O)
67	GND	PAR64	GND	PAR64

Table 13.9 PCI Local Bus Configuration *(continued)*

| Pin | 5-V system environment | | 3.3-V system environment | |
---	Side B	Side A	Side B	Side A
68	AD[63]	AD[63]	AD[62]	AD[63]
69	AD[61]	GND]	AD[61]	GND
70	+5 V (I/O}	AD[60]	+3.3 V (I/O)	AD[60]
71	AD[59]	AD[58]	AD[59]	AD[58]
72	AD[57]	GND	AD[57]	GND
73	GND	AD[56]	GND	AD[56]
74	AD[55]	AD[54]	AD[55]	AD[54]
75	AD[53]	+5V (I/O)	AD[53]	+3.3V (I/O)
76	GND	AD[52]	GND	AD[52]
77	AD[51]	AD[50]	AD[51]	AD[52]
78	AD[49]	GND	AD[40]	GND
79	+5V (I/O)	AD[48]	+3.3V (I/O)	AD[48]
80	AD[47]	AD[46]	AD[47]	AD[46]
81	AD[45]	GND	AD[45]	GND
82	GND	AD[44]	GND	AD[44]
83	AD[43]	AD[42]	AD[43]	AD[42]
84	AD[41]	+5V (I/O)	AD[41]	+3.3V (I/O)
85	GND	AD[40]	GND	AD[40]
86	AD[39]	AD[38]	AD[39]	AD[38]
87	AD[37]	GND	AD[37]	GND
88	+5 V (I/O)	AD[36]	+3.3V (I/O)	AD[36]
89	AD[35]	AD[34]	AD[35]	AD[34]
90	AD[33]	GND	AD[33]	GND
91	GND	AD[32]	GND	AD[32]
92	Reserved	Reserved	Reserved	Reserved
93	Reserved	GND	Reserved	GND
94	GND	Reserved	GND	Reserved[**]

[*] Connector key = 3.3 V.
[†] Connector key = 5.0 V.
[‡] Connector key = 64-bit spacer.
[§] 64-bit start.
[**] 64-bit end.

CHAPTER 14

Memories

INTRODUCTION

Automatic mechatronic devices usually are designed to perform a sequence of tasks that have been previously programmed and stored in memory. The type of memory that the mechatronics designer can use in a given project depends on the following considerations:

- Required capacity

- Volatile or nonvolatile operation

- Required speed

If the mechatronic project is connected to a computer, the computer's memory can be used for data and program storage. But if the project is designed for autonomous operation, we have two possibilities:

1. The device can use its own microprocessor and a large enough memory to store data and programming.

2. The device is too simple to require a microprocessor but needs enough memory to store a short sequence of operations or tasks to be done.

In this chapter, we provide some solutions to the problem of data and program storage in mechatronic designs.

SIMPLE MEMORY TYPES

Terminal strips or slots can be used as simple programmed memories as shown in Fig. 14.1. The positions of the jumpers or interconnections make up the stored data.

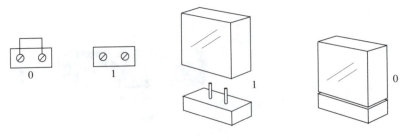

Figure 14.1

Figure 14.2 shows a printed circuit board with copper lines defining the stored data. This printed circuit board acts as a memory card when inserted in a slot as shown in Fig. 14.2.

Figure 14.3 shows a simple optical memory card with a read circuit. This circuit puts a "0" in the position that corresponds to a hole and a "1" in the position with no holes.

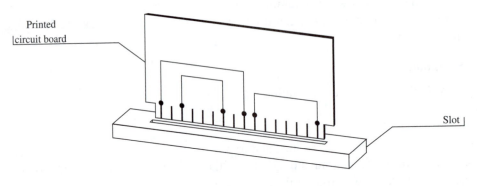

Figure 14.2

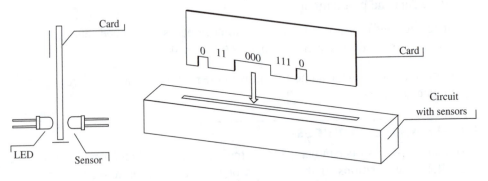

Figure 14.3

DIODE MATRIX MEMORY

Simple command sequences can be generated using a card with diodes and a sequencer such as the 4017, as shown in Fig. 14.4. This circuit produces a sequence of ten commands, generated by the position of the diodes in the card, as shown in the figure.

The output of the circuit is CMOS compatible and can control many of the drive stages shown in the previous parts of this book (see Chapters 4, 5, and 8). The output stages used with a parallel port (Chapter 13) can also be used with these cards.

Figure 14.5 shows how two 4017s can be cascaded to produce sequences of up to 19 pulses. The diode matrix card can be installed on a small printed circuit board and plugged into a slot in the mechatronic project. This way, changing the programming is very easy; you simply replace the card with another one that contains a new command sequence.

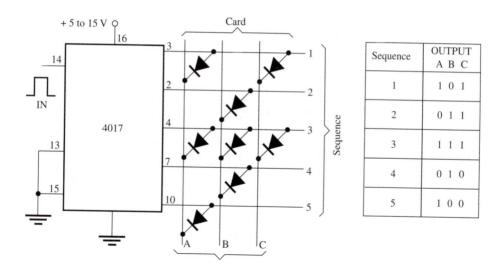

Sequence	OUTPUT A B C
1	1 0 1
2	0 1 1
3	1 1 1
4	0 1 0
5	1 0 0

Figure 14.4

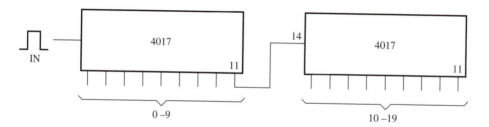

Figure 14.5

MECHANICAL MEMORY

A simple mechanical device for storing small amounts of data is shown in Fig. 14.6. Shown are methods that use (a) contacts and (b) magnets. In Fig. 14.6a, the card engages selected contacts when it is inserted into a slot. In Fig. 14.6b, magnets determine which reed switch is closed when the card is inserted.

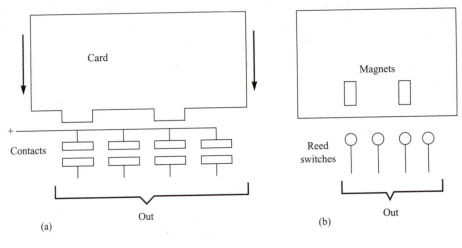

Figure 14.6

ANALOG (SAMPLE AND HOLD) MEMORY

Analog data (in the form of voltage from transducers) can be stored for periods ranging from a fraction of a second to several seconds using sample-and-hold circuits as shown in Fig. 14.7. When the FET is turned on by the application of a positive voltage applied to its gate, the voltage present in the input is transferred to capacitor C. When the transistor switches off, the capacitor maintains the voltage level, which is read to the operational amplifier, which presents it at its output.

Since the input impedance of the operational amplifier is very high, the voltage is kept constant over a large time interval. The low impedance of the operational amplifier allows the use of the stored voltage to control external devices (e.g., pulse-width modulated, linear, or other controls).

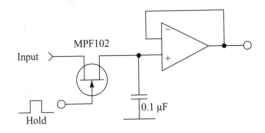

Figure 14.7

SEMICONDUCTOR MEMORIES

When a project requires more extensive storage for data and programming, it is appropriate to use semiconductor memories. Many types can be used in such projects.

Memories are broadly classified as being volatile or nonvolatile. Volatile memories lose all of their stored data when the power supply is switched off. Conversely, non-volatile memories can maintain the stored data even their power source is switched off.

Flip-flops can be used to store data. Each flip-flop, as shown in Fig. 14.8, can be used to store one bit. A shift register also can be used to store data as shown in Fig. 14.9.

There are also ICs that are built specifically to work as memories. Some general types are as follows.

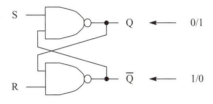

Figure 14.8

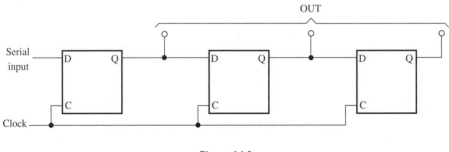

Figure 14.9

RAM

There are two types of random access memory (RAM): static and dynamic. The static RAM is a volatile memory in which data can be read or written any time. It stores data until the moment the power supply is disconnected. Figure 14.10 shows an application of the 2102, a simple static RAM that has 32 rows of 32 cells for a total storage of 1024 bits of data. The 2102 is dated today, but it is excellent for instructional purposes in simple mechatronic projects.

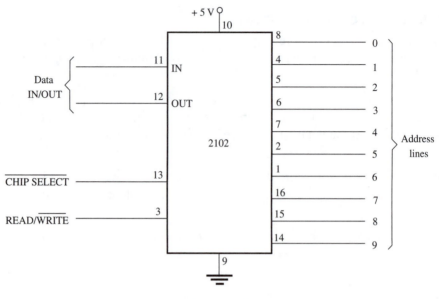

Figure 14.10

This memory is TTL, so it must be powered from a 5-V supply. It can drain or deliver several milliamperes at the outputs.

Other common RAM memories are the 2114 (1024 × 4 bits) and the 4016 (2k × 8 bits). Figure 14.11 shows a circuit that uses the 2114.

Dynamic RAM stores the data while the power supply is on, but it must be "refreshed" by an additional signal. The refresh signal recharges the small capacitors in which the bits are stored.

A popular RAM (also dated today, but interesting for simple instructional projects) is the 2116. Figure 14.12 shows an application circuit. This memory is organized in a 16 k × 1 array of bits.

Using the 2116

a. Two sets of addresses are required. First, we provide 7n column addresses and then provide seven row addresses to the same pins.

b. Output data is available only after address selection is completed. Data output must be latched externally.

c. Refresh is needed. Each address row must be addressed at least 500 times per second.

ROM

Read-only memories (ROMs) are programmed as part of the manufacturing process. Stored data cannot be changed; it can only be read.

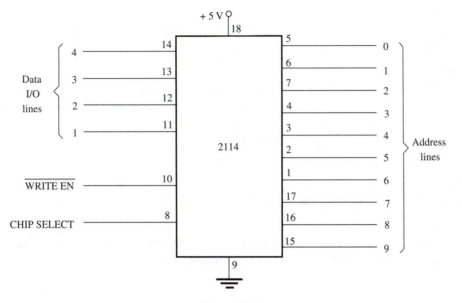

Figure 14.11

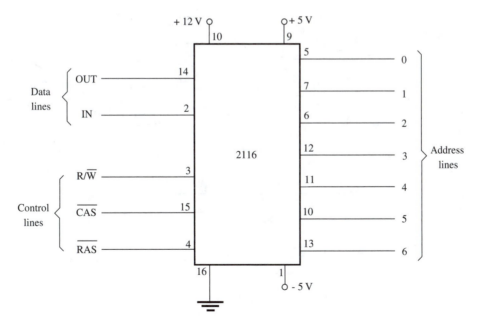

Figure 14.12

PROM

Programmable read-only memories (PROMs) consist of an array of fuses that are burned to create programming. Thus, the PROM can be programmed only once.

EPROM

Erasable programmable read-only memories (EPROMs) are programmed byte by byte using special circuitry or using a computer and programming software, as suggested by Fig. 14.13. After EPROMs are programmed, data can only be read (not modified by electrical signals). However, they can be erased by intense ultraviolet light and reprogrammed as shown in Fig. 14.14.

There are also one-time programmable (OTP) EPROMs, also called one-time programmable read-only memories (OTPROMs). To reduce costs, these devices lack the option of erasure.

EPROMs with capacities ranging from several kilobytes to many megabytes are available and are ideal for use in mechatronic projects.

Popular types are as follows:

- 2716—16 kb (1k × 8)
- 2732—32 kb (2k × 8)
- 27C1024—1 Mb (128k × 8)

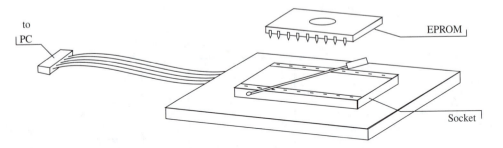

Figure 14.13

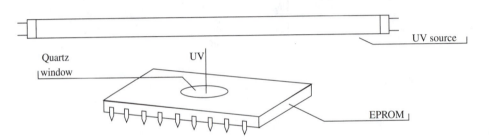

Figure 14.14

Figure 14.15 shows the pinout of the 2716. Table 14.1 shows typical EPROM programming voltages as specified by the most popular memory manufacturers.

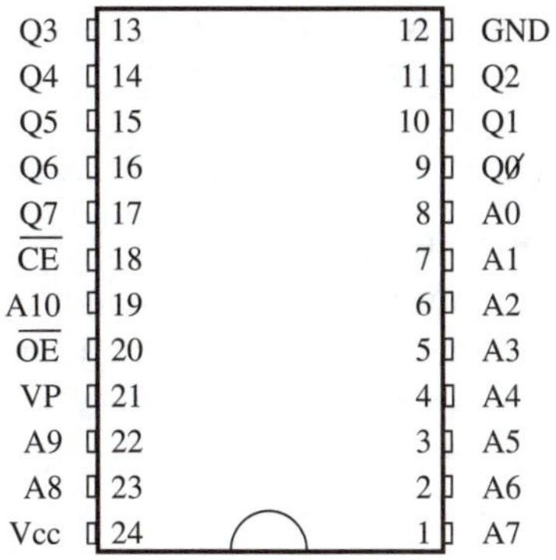

```
      Q3  ⊏ 13          12 ⊐ GND
      Q4  ⊏ 14          11 ⊐ Q2
      Q5  ⊏ 15          10 ⊐ Q1
      Q6  ⊏ 16           9 ⊐ Q0̸
      Q7  ⊏ 17           8 ⊐ A0
      CE  ⊏ 18           7 ⊐ A1
     A10  ⊏ 19           6 ⊐ A2
      OE  ⊏ 20           5 ⊐ A3
      VP  ⊏ 21           4 ⊐ A4
      A9  ⊏ 22           3 ⊐ A5
      A8  ⊏ 23           2 ⊐ A6
     Vcc  ⊏ 24           1 ⊐ A7
```

Figure 14.15

Table 14.1   Typical EPROM Programming Voltages

Type	Programming voltage (Vpp)
2716	25
27C16	25
2732	25
2732A	21
27C32	21
2764	21
2764A	12.5
27C64	21
27128D	21
2816	5
2864A	5

EEPROM

Electrically erasable programmable read-only memories (EEPROMs) are similar to EPROMs except that they can be erased by an electrical signal rather than a UV light source. A special type of EPROM is the flash EEPROM. These are similar to standard EPROMs except that they can be erased in one operation, whereas standard EEPROMs must be erased one byte at a time.

EPROM and EEPROM numbers can be decoded as shown in Table 14.2.

Table 14.2   EPROM and EEPROM Number Codes

Number	Type
27(C)XXX	EPROM or OTPROM
57(C)XXX	EPROM or OTPROM with multiplexed data lines
28 (C) XXX	EEPROM, with the C indicating CMOS
28(F)XXX	Flash EEPROM

USING SEMICONDUCTOR MEMORIES

A typical application for a memory in a mechatronic project is shown by the blocks in Fig. 14.16. A clock circuit determines the execution timing of each command, which is stored in one line of memory. The memory stores the entire sequence of commands to be executed. The output of each memory line controls one of the external devices through the use of a power stage such as the ones shown in Chapters 4, 5, 6, and 8.

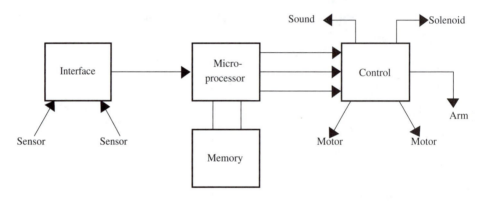

Figure 14.16

CHAPTER 15

Programmable Controls

PROGRAMMED AND SEQUENTIAL MECHANICAL SENSORS

Using simple arrangements of wheels and screws, we can create position sensors or programmed sensors such as shown in Fig. 15.1. Devices shown are as follows:

Figure 15.1a. In this figure, a metal-coated cylinder is used as a position sensor. The blades in the array touch the metal only at the points where there is no coating of insulation. The designer can arrange the coated and uncoated positions of the cylinder to provide the proper switching function.

Figure 15.1b. Here, a worm screw is used to create a programmed, sequential switch.

Figure 15.1c. A plastic or printed circuit disc coupled to a gear box or a clock mechanism can be used in a sequential control as shown here. The speed of the disc determines the duration of the control cycle.

ENCODERS

Encoders (optical or mechanical) are transducers that convert angular or linear motion into digital signals. They can be coupled to shafts of motors, mechanisms, or any other mechanical part of a project to sense and control its position or motion.

There are two types of encoders, as described below.

1. *Incremental rotary encoders.* Incremental encoders are used to measure rotational movement. They can be used to determine angular speed or angular position. This kind of encoder registers positions or speed simply by counting the number of pulses generated by the rotary disc.

2. *Absolute rotary encoders.* With this kind of encoder, angular position can be read from digital information on the rotary disc. The position of a device is

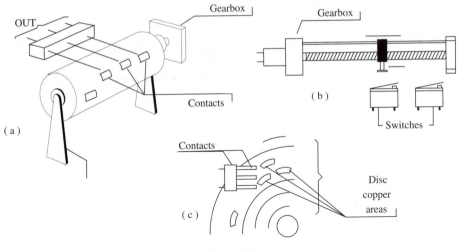

Figure 15.1

available as a digital number as soon as the equipment is powered on. Normally, the digital positional values are represented in Gray Code. In this code, only one bit changes as we pass from one value to the next (see Table 15.1).

Table 15.1   Decimal vs. Binary vs. Gray Code

Decimal	Binary	Gray	Decimal	Binary	Gray
0	0000	0000	8	1000	1100
1	0001	0001	9	1001	1101
2	0010	0011	10	1010	1111
3	0011	0010	11	1011	1110
4	0100	0110	12	1100	1010
5	0101	0111	13	1101	1011
6	0110	0101	14	1110	1001
7	0111	0100	15	1111	1001

Figure 15.2 shows the optical discs corresponding to the two types of encoders.

Mechanical encoders use a conductive plastic circuit component and a multiple-finger collector system (the same principle shown in Fig. 15.1). The typical resolution is 100 pulses/360°. The output signal should be debounced using an appropriate circuit.

Optical encoders use a disc with dark and white areas to represent the digits to be read, or to produce pulses when rotating, as shown in Fig. 15.2. There are many ways to read the information from a disc to detect position or speed. Several are illustrated in Fig. 15.3, including (Fig. 15.3a) a transparent disc, (Fig. 15.3b) a

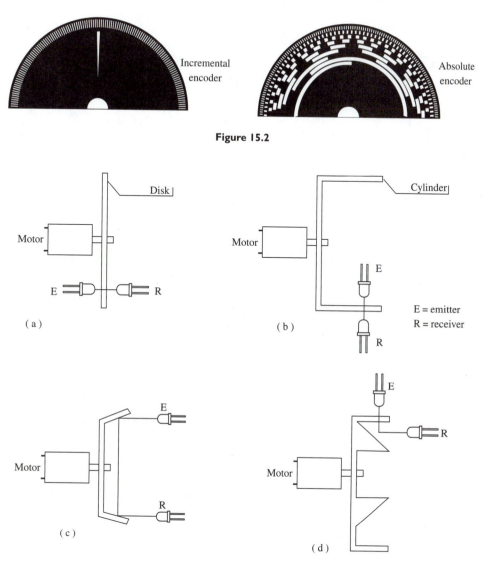

Figure 15.2

Figure 15.3

transparent cylinder, and (Fig. 15.3c and 15.3d) reflection of light from a source to a sensor. Table 15.2 illustrates encoder resolution by power of two and angle.

CIRCUITS

Figure 15.4 shows some circuits used to read the speed or position of optical encoders. Here, we have circuits based on (Fig. 15.4a) a bipolar transistor, (Fig. 15.4b) a comparator, and (Fig. 15.4c) a logic trigger. The light source can be an infrared or visible LED, and the sensor can be a phototransistor. These circuits can be used to drive logic stages (TTL or CMOS) or other circuitry suggested in this book.

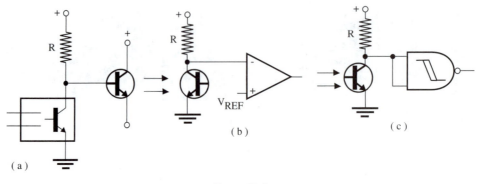

(a)

(b)

(c)

Figure 15.4

Table 15.2   Resolution of an Encoder by Power of Two and Degrees

No. of bits	Power of 2	Angle	No. of bits	Power of 2	Angle
1	2	180°	13	8,192	2.64'
2	4	90°	14	16,384	1.32'
3	8	45°	15	32,768	39.55"
4	16	22.5°	16	65,536	19.78"
5	32	11.25°	17	131,072	9.89"
6	64	5.63°	18	262,144	4.95"
7	128	2.81°	19	524,288	2.47"
8	256	1.41°	20	1,048,576	1.24"
9	512	42.19'	21	2,097,152	0.62"
10	1,024	21.09'	22	4,194,304	0.31"
11	2,048	10.55'	23	8,388,608	0.15"
12	4,096	5.27'	24	16,777,216	0.08"

PLCS

The programmable logic controller (PLC) device is widely used in industry, mechatronics, and other fields to automate some basic functions. For instance, in a factory, the PLC can be used with sensors in photoelectric applications such as bar code scanners and to control relays, motors, and other accessories for many purposes (see Fig. 15.5).

Using appropriate sensors, a PLC can control most routine processes. If you can create a flow chart and state diagram of the process to be controlled, then it is very likely that you can program a PLC to perform the task.

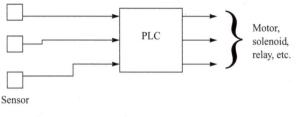

Figure 15.5

A PLC can use various kinds of input/output (I/O) cards to interface with sensors and actuators. The card can use 5-V or 110-V sources as input or output. The PLC can add intelligent control to a mechatronic device.

SOFTWARE

A PLC uses a command structure called *ladder logic*. When a line goes to the high output level, the controls connected to it are activated. Which line goes to the high output level depends on the program and sensors that have been activated.

Typical PLCs can store data and programs ranging from several kilobytes to many megabytes, and they can control many outputs using signals picked up by many sensors.

Some representative PLC manufacturers are as follows:

ABB	Alfa Laval
Allen-Bradley	ALSTOM/Cegelec
Aromat	AutomationDirect/PLC Direct/Koyo/
B&R Industrial Automation	Berthel gmbh
Cegelec/ALSTOM	Control Microsystems
Crouzet Automatismes	Control Technology Corporation
Cutler Hammer/IDT	Divelbiss
EBERLE gmbh	Elsag Bailey
Entertron	Festo/Beck Electronic
Fisher & Paykel	Fuji Electric
GE-Fanuc	Gould/Modicon
Grayhill	Groupe Schneider
Hima	Hitachi
Honeywell	Horner Electric
Idec	IDT/Cutler Hammer
Jetter gmbh	Keyence
Kirchner Soft	Klockner-Moeller
Koyo/AutomationDirect/PLC Direct	Microconsultants
Mitsubishi	Modicon/Gould
Moore Products	Omron

Opto22	Pilz
PLC Direct/Koyo/AutomationDirect	Reliance
Rockwell Automation	Rockwell Software
SAIA-Burgess	Schleicher
Schneider Automation	Siemens
Sigmatek	SoftPLC/Tele-Denken
Square D	Tele-Denken/SoftPLC
Telemecanique	Toshiba
Triangle Research	

Many PLCs used in industrial controls can be adapted to all levels of mechatronic and robotic projects.

Logic

INTRODUCTION

The basic idea of logic circuitry comes from digital electronics. It is important to understand some fundamentals of digital electronics so as to better comprehend how devices in this important group function. Digital electronics is based on the idea that any quantity can be represented by zeroes and ones.

Using only two digits, we can represent any quantity in the *binary* numbering system. Figure 16.1 shows how a number in the decimal system can be converted to binary. The advantage of using binary code is that it can be handled more easily by electronic circuits.

In its simplest form, an open switch can be used to represent a 0, and a closed switch can be used to represent a 1, as shown in Fig. 16.2. Working with open and closed switches, it is possible to perform any mathematical or logic operations with binary numbers. The rules that command these operations were first established by Robert Boole two centuries ago; hence the term *Boolean logic*.

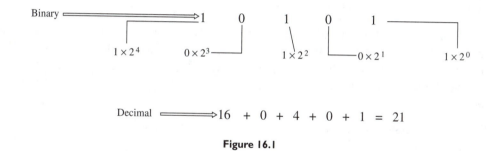

Figure 16.1

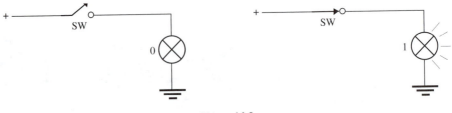

Figure 16.2

Moving now to the idea that a transistor (like some other components) can also act as a switch, we conclude that it can be used not only to represent a zero or a one but also to execute operations using binary numbers. As shown in Fig. 16.3, if a transistor is switched to its saturated (conducting) state, the voltage in its collector falls to zero, or a value near zero, allowing it to represent a low logic level or 0.

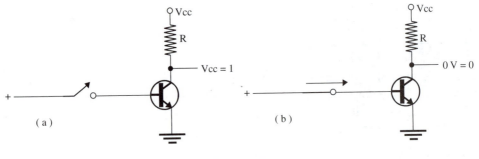

(a) (b)

Figure 16.3

On other hand, if the transistor is switched off (nonconducting), the voltage in its collector rises to a level near the power supply value. This voltage can be used to represent a high logic level or a 1. Using many transistors in parallel, we can represent any complete number; by applying appropriate command signals to them, we can execute mathematical or logical operations.

A single digit, one or zero, that is used to represent a binary number, is called a *bit* (a contraction of *binary digit*), and a group of eight bits is called *byte*. A group of four bits is also called *nibble*. A byte can be used to represent any quantity or decimal number between 0 (00000000) and 255 (11111111).

Thousands or even millions of transistors wired as switches can be placed on a chip in configurations that allow them to perform operations with binary numbers. The numbers are processed as bits, which are assembled as the final bytes. Such chips are the digital circuits as found in computers, microprocessors, microcontrollers, and so on.

Several basic functions are used in projects that employ digital circuits. These functions are represented in Fig. 16.4.

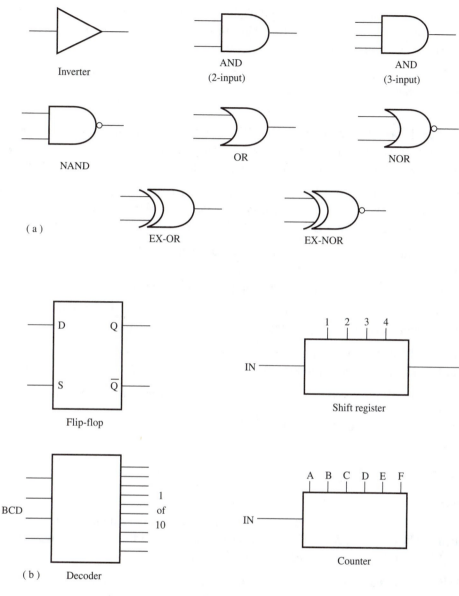

Figure 16.4

In the first group of basic functions, we find devices called *gates*. Gates are circuits that combine two or more logic levels or digital signals to produce a resulting logic level or signal as determined by their internal logic rules. The rules follow the mathematics created by Boole. In the second group, we have devices that perform more complex functions, such as flip-flops (which can store a bit), multiplexers, counters, decoders, and other devices.

These circuits form *families* of digital integrated circuits, and all devices in each family have common characteristics. This allows them to be interconnected directly to form more complex devices.

Modern equipment often uses integrated circuits from two main families, as described below.

TTL

The transistor-transistor logic (TTL) IC family consists of many devices (nearly 1000) with several common characteristics. They are all powered from a 5-V power supply, and the output of one can be connected directly to the input of another.

The standard family of TTL devices comprises a large number of functions and includes such devices as gates, flip-flops, counters, decoders, multiplexers, demultiplexers (MUXs and DEMUXs), phase-locked loops (PLLs), timers, and so forth.

Some subfamilies of this group of ICs have special characteristics, e.g., the low-power Schottky (LS). This group is compatible with circuits found in computers, adding the features of low power consumption and high speed.

TTL devices are easily recognized by the designation "74" at the beginning of their part numbers. Sometimes, the TTL family is referred to as the 74 family or 74xx.

Almost all of devices in this family are designated by a 74 followed by another number that defines their function. Examples are 7400, 7492, 7191, and so on.

The subfamilies are recognized by one or two letters after the 74. For example,

- 74LS04 (low-power Schottky)
- 74H93 (high-power device)
- 74S121 (Schottky)

CMOS

The other important logic family consists of complementary metal-oxide semiconductor (CMOS) integrated circuits. These circuits perform the same functions as the TTL, but they have different electric characteristics.

CMOS devices can be powered from supply voltages ranging from 3 to 15 V or 3 to 18 V. They are not as fast as TTL, but their power consumption is lower, and their input impedance is very high.

The devices in this family are designated by a "40," followed by a number that indicates the device's function. We encounter designations such as 4004, 4017, 4093, 40121, and so on.

Some CMOS subfamilies use the same designation or number as the equivalent TTL function. These devices are indicated by the 74 code but with a C to signify that they are CMOS. An example is the 74C00.

Specifications

The function of each digital integrated circuit is indicated by its part number. The basic electrical characteristics are the same throughout the family. It is important to remember that TTLs always need a 5-V power source, but CMOS can use any voltage between 3 and 15 V.

Warning

Integrated circuits of the CMOS family are very sensitive to electrical charges that may be stored in your body in the form of static electricity, and CMOS transistors can be damaged by electrostatic discharge (ESD). So, *never touch the leads of these devices directly.* ESD can destroy the device.

IMPLEMENTATION OF LOGIC DEVICES

As with other devices, the configuration of CMOS and TTL devices depends on several factors. In modern equipment, the tendency is to place all circuit functions on a single IC. As a result, one chip can contain the equivalent of hundreds or even thousands of TTL or CMOS ICs. Although this does not change the operating principles, this approach means that you cannot access circuits individually. If one of them fails, you must replace the entire IC.

LOGIC TESTING

Digital ICs generally must be tested using special instruments. The simplest way is to measure the relevant voltages with a multimeter, but this does not always provide a conclusive result (depending how the circuit operates). An advanced test uses logic probes or oscilloscopes to observe the digital signals produced by each IC.

LOGIC CONVENTIONS

We usually use a "1" to represent the high logic level or "+V" (the power supply voltage), and we use a "0" to represent the low logic level (ground[*]).

Digital logic may be divided in two classes:

1. In *combinational logic,* the outputs are determined by the logical operation being performed and the logic states at the inputs at that moment.

2. In *sequential logic,* the outputs depend on either of the logical inputs and the previous states of those outputs.

Combinational logic devices include inverters, AND gates, NAND gates, OR gates, NOR gates, exclusive OR gates, binary addition blocks, multiplexers, and decoders/demultiplexers.

Sequential logic devices include RS NAND latches, clocked RS latches, RS flip-flops, JK flip-flops, and D latches.[†]

---

[*]In digital electronics, "ground" does not represent an Earth connection as used for safety in AC power electronics, home electrical wiring, and so forth. It refers only to the relative ground (0-V) of the circuit.

Other Conventions

Vdd	The positive power supply
Vss	Ground (0-V reference)
Vee	Bias voltage
A, B, C, D, etc.	Logic block inputs
X, Y, W	Logic block outputs
Q	Output of flip-flops and bistable elements
/Q	Complementary output of bistable elements
CLK	Clock
CLR	Clear

THE CMOS AND TTL FAMILIES

As discussed previously, the two main logic families upon which most modern digital electronic projects are based are transistor-transistor logic (TTL) and complementary metal-oxide semiconductor (CMOS) devices. The two logic families have different capabilities, limitations, advantages, and disadvantages. TTL technology, created by Texas Instruments, remained the top choice for all designers until a new technology appeared.

In 1960, the technology known as CMOS was created as an attempt to replace the rival TTL technology. CMOS ICs are easy to use, cheap, have a high input impedance, and consume near zero power. Another advantage of CMOS over TTL is the ability to function within a wider voltage range (3 to 15 V) as compared with the fixed voltage of 5 V for TTL. Circuits designed to use this family are therefore more flexible, as they can be powered from 3 to 15 V supplies.

The low current consumption CMOS devices makes them ideal for applications that use battery power. On the other hand, the low speed of CMOS devices means that the TTL family is a better choice for desktop applications where high operational speed is needed.

CMOS ICs are found in two versions: 4000A (standard) and 4000B (buffered). Some characteristics of each are listed below.

a. 4000A

  – Released in 1972
  – 3- to 12-V operation

†RS refers to the device's reset/set input designators. In a JK flip-flop, the inputs have been renamed J and K, in honor of Jack Kilby, the Texas Instruments engineer who invented the integrated circuit in 1958. The "D" in D latch stands for "data."

– Poor output symmetry
– Outputs sensitive to input signals
– Current consumption directly proportional to switching frequency

b. 4000B

– Released in 1975
– Use of "buffered" inverters in series, increasing linearity
– Good output symmetry
– Longer propagation delay than series A
– Current consumption directly proportional to switching frequency

c. Another series, the 4000UB, contains unbuffered circuits, increasing speed. Since their gain is only 23 dB, they are ideal for analog applications.

Other Subfamilies

Many CMOS subfamilies have been created since 1972, and each one has special characteristics that can be useful for specific applications. In particular, the 74 series is worthy of consideration. The idea behind this series is to provide the designer with CMOS devices that are equivalent in function and package to devices in the 7400 TTL family.

Many CMOS 74-series subfamilies have been created. The main ones are as listed below.

a. *Standard (74C00).* This family used normal MOSFET-type CMOS technology and is now obsolete.

b. *High-speed (74HC00).* Introduced in 1980, this technology offers the speed of regular TTL with CMOS characteristics.

c. *High-speed (74HCT00).* Here, the inputs are compatible with TTL outputs.

d. *Advanced high-speed (74AC00).* These devices have typical propagation delays of 5 ns.

e. *Advanced high-speed (74HCT00).* The series has TTL-compatible inputs. Propagation delays of 7 ns are typical.

Table 16.1 provides various characteristics of the primary CMOS and TTL families for comparison purposes.

CMOS CHARACTERISTICS

The characteristics of CMOS ICs are determined by the MOS transistors inside them. Although CMOS ICs are designed for digital operation, in some cases they can be used as analog amplifiers.

Table 16.1   CMOS and TTL Characteristics

	Supply voltage range	Quiescent current (per gate)	Propagation delay (per gate)	Maximum operating frequency	Fan-out (to LS TTL inputs)
4000B	3–15 V	0.01 μA	125 ns (5 V) 50 ns (10 V) 40 ns (15 V)	2 MHz (5 V) 5 MHz (10V) 8 MHz (15 V)	1
4000UB	3–15 V	0.01 μA	90 ns(5V) 50 ns (10V) 40 ns (15 V)	3 MHz (5 V) 5 MHz (10V) 8 MHz (15V)	1
74HC00	2–6 V	0.02 μA	8 ns	40 MHz	10
74HCT00	4.5–5.5 V	0.02 μA	10 ns	–	10
74AC00	2–6 V	0.02 μA	5 ns	100 MHz	10
74ACT	4.5–5.5 V	0.02 μA	7 ns	–	60
Regular TTL	4.5–5.5 V	2 mA	10 ns	35 MHz	20
High-power TTL	4.5–5.5 V	4.1 mA	6 ns	50 MHz	40
Low-power TTL	4.5–5.5 V	0.2 mA	33 ns	3 MHz	5
Schottky TTL	4.4–5.5 V	4 mA	3 ns	125 MHz	40
TTL LS	4.75–5.25	0.5 mA	9 ns	40 MHz	20

a. *Linear operation.*   The voltage transfer characteristic of a CMOS inverter is shown in Fig. 16.5. In this figure, we can see that, if we bias the input in the linear region of the curve, the device can be used as an analog amplifier.

b. *Digital operation.*   The current drained or delivered by any CMOS output depends on the power supply voltage according to Table 16.2.

Table 16.2   Voltage and Output of CMOS ICs

Power supply voltage	Output current (delivered or drained)
5.0 V	0.88 mA
10.0 V	2.2 mA
15.0 V	8.0 mA

In some ICs, the outputs are not symmetric. In these cases, the maximum drained current is different, and the output characteristic of the device varies according to the specific type.

USING CMOS DEVICES

Some important considerations in the use of CMOS devices are described below.

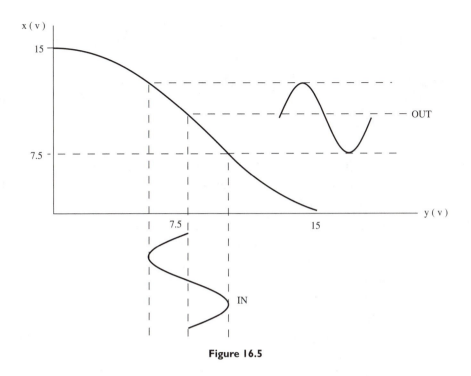

Figure 16.5

Electrostatic Protection

CMOS ICs use insulated gate field-effect transistors (IGFETs), which have nearly infinite input impedance. This means that, if excess voltage is applied to any input or output of a CMOS circuit, it can break through the gate insulation of the transistors, destroying the device.

The human body, under certain conditions, can store electrical charges at voltages high enough to destroy a CMOS element.[*] Although ICs have internal elements designed to protect them against such high voltages, their protective functions are limited. Laboratory tests indicate that they can survive spikes reaching only a few kilovolts. Therefore, when handling the CMOS ICs, you must never touch the leads. Avoid nylon clothing, and don't work in places where mats or carpets can create and store electric charges. It is a good idea to use a grounded metal wrist strap when working with CMOS ICs.

Unused Inputs

Any inputs not used in a particular project must not be left disconnected or *floating*. The input impedance of these leads can make them sensitive and able to pick up

[*]Your body can generate and hold charges up to 35,000 V when walking on synthetic carpet at low (20%) relative humidity.

signal spikes that affect logic functions. A second problem with floating an input is that it can push the transistor inside the IC into the linear region of the operational curve, where the power consumption is higher.

The unused inputs must be tied to ground or power supply voltage (Vcc), depending on the function of the gate. Figure 16.6a shows how the unused inputs can be connected.

Decoupling

The fast logic level changes of a CMOS ICs can cause current spikes to propagate throughout the circuit. If the sensitive elements of the circuit are not decoupled, these spikes can affect their operation, causing them to go to unintended states.

When designing a printed circuit board for CMOS applications, the designer must be careful with traces that deliver power to the ICs and, when necessary, must decouple these devices. A ceramic capacitor (0.1 µF) wired between the positive power supply voltage (Vdd) and ground (Vss) of a CMOS IC normally is sufficient to decouple the devices, as shown in Fig. 16.6b.

POWER SUPPLIES FOR CMOS

CMOS devices have a very low quiescent current, making them ideal for battery-powered applications. However, it sometimes is necessary to drive them from the AC power line. Since the source voltage is not critical, regulated power supplies are not always needed.

Figure 16.7 shows some simple circuits that can be used to power CMOS projects from the AC power line. These circuits are described in the following text.

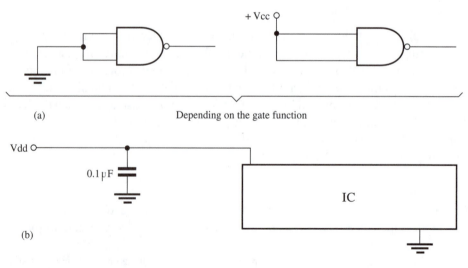

(a)   Depending on the gate function

(b)

Figure 16.6

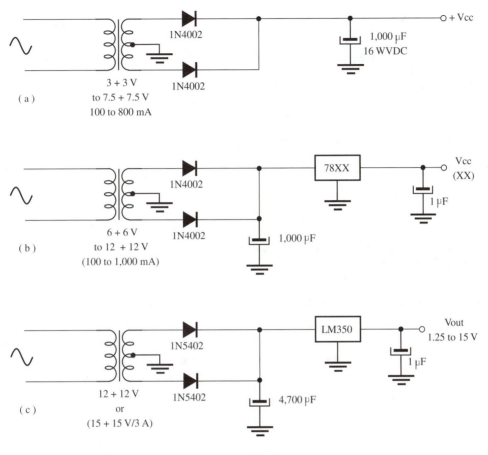

Figure 16.7

Figure 16.7a.   This circuit is the simplest and can supply projects that are not sensitive to voltage variations. The voltage is unregulated, and the current depends on the transformer's secondary. Transformers with windings ranging from 3 to 9 V can be used.

Figure 16.7b.   This is a regulated power supply for currents up to 1 A (depending only on the secondary winding of the transformer). The voltage is determined by the IC. The XX determines the voltage. For instance, the 7809 delivers 9 V to the output. The IC must be mounted on a heatsink.

Figure 16.7c.   Finally, for applications in which the designer wants to change the voltages, this circuit is recommended. The output can be adjusted from 1.25 to 15 V, and the output current is up to 3 A. The IC must be mounted on a heatsink.

TTL CHARACTERISTICS

Regular or standard TTLs can deliver and sink only few milliamperes per output, so they need a drive stage to control medium- and high-power loads.

Typical output currents are as follows:

- HI-level output current (source): 0.4 mA (typ.)

- LO-level output current (drain): 16 mA (typ.)

Other general characteristics are as follows:

- LO logic level: 0 to 0.8 V

- HI logic level: 2.4 to 5 V

INTERFACING

CMOS integrated circuits in the 4000 family cannot deliver or drain high currents at their outputs. The maximum current you can get for any device in this family is only enough to drive small loads such as LEDs and other low-power devices.

Typically, the current that can be driven or delivered by any CMOS output remains below several milliamperes and is dependent on the power supply voltage. Table 16.3 gives the current delivered or drained by the outputs as function of the power supply voltage.

Table 16.3   Power Supply Voltage vs. Output Current

Power supply voltage	Output current (delivered or drained)
5.0 V	0.88 mA
10.0 V	2.2 mA
15.0 V	8.0 mA

If you need to drive loads requiring higher current than CMOS ICs can deliver or drain, an intermediate power amplification stage is needed. Depending on the application, many possible configurations can be chosen. Several are suggested in the following paragraphs.

We begin with low-speed configurations that can be used to turn loads on and off using the logic levels found at the output of CMOS ICs. To drive loads with such high-speed signals as audio and RF, other configurations will be suggested.

It is also important to remind the reader to use protection for the driving devices if inductive loads as relays, solenoids, DC motors, or others are controlled by the stages. Figure 16.8 shows the simplest way to protect devices (e.g., bipolar transistors, power FETs, SCRs, and others) against voltage spikes created by fast current

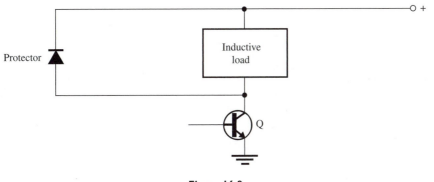

Figure 16.8

changes in the circuit. Any general-purpose or silicon diode can be used for this task, such as the 1N914, 1N4148, or 1N4002.

ELECTROMAGNETIC INTERFERENCE

Fast current changes across a circuit, particularly if it is inductive, can cause electromagnetic interference (EMI). High-frequency pulses can interfere with radio receivers and other devices that operate with electromagnetic waves and are located near the circuit. The high-voltage spikes generated in this process can also cause a circuit to go to an unstable state (see *Decoupling*, p. 344).

PRACTICAL INTERFACE CIRCUITS

Figure 16.9 shows some practical circuits for interfacing CMOS and TTL logic, with power stages and loads. They are described below.

Figure 16.9a.   Medium-power stage using a Darlington NPN transistor

This circuit is used if your application needs more than 100 mA to be driven. This block uses an NPN Darlington transistor to supply the load with currents in the range of several amperes to more than 5 A. The load current depends on the transistor used.

Transistor	Current (I_c)	Voltage (V_{ce})
TIP110	1.25 A	60 V
TIP111	1.25 A	80 V
TIP112	1.25 A	100 V
TIP120	3 A	60 V
TIP121	3 A	80 V
TIP122	3 A	100 V
TIP140	5 A	60 V
TIP141	5 A	80 V
TIP142	5 A	100 V

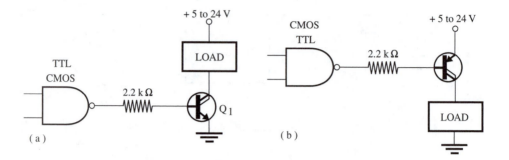

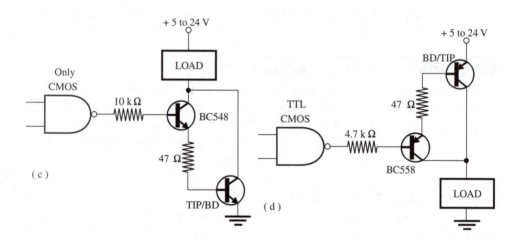

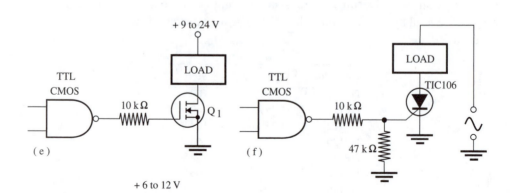

Figure 16.9   *(continues)*

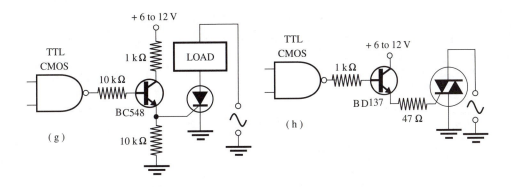

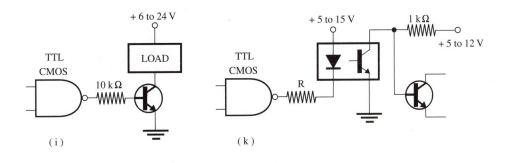

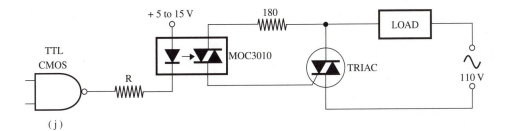

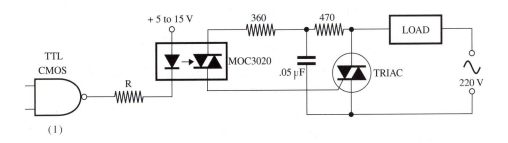

Figure 16.9 *(continued)*

Figure 16.9b.   Medium-power stage using a Darlington PNP transistor

The equivalent configuration using a PNP Darlington transistor is shown here. The suitable transistors are given in the table, and the load is on when the output of the IC goes to the LOW logic level. Remember to install the transistor on a heatsink.

Transistor	Current (Ic)	Voltage (Vce)
TIP115	2 A	60 V
TIP116	2 A	80 V
TIP117	2 A	100 V
TIP125	3 A	60 V
TIP126	3 A	80 V
TIP127	3 A	100 V
TIP145	5 A	60 V
TIP146	5 A	80 V
TIP147	5 A	100 V

Figure 16.9c.   Medium-power stage using discrete Darlington NPN transistors

A discrete bipolar transistor can be wired to form a Darlington stage. The high current gain of this stage is suitable for applications as driving high-current loads from the low-current output of logic circuits.

Transistor	Current (Ic)	Voltage (Vce)
BD135	1 A	45 V
BD137	1 A	60 V
BD139	1 A	80 V
TIP31	3 A	40 V
TIP31A	3 A	60 V
TIP31B	3 A	80 V
TIP31C	3 A	100 V
TIP41	6 A	40 V
TIP41A	6 A	60 V
TIP41B	6 A	80 V
TIP41C	6 A	100 V

Figure 16.9d.   Medium-power stage using discrete Darlington PNP transistors

The equivalent configuration using PNP discrete transistors to form a Darlington stage is given here.

Transistor	Current (Ic)	Voltage (Vce)
BD136	1 A	45 V
BD138	1 A	60 V
BD140	1 A	80 V

Transistor	Current (Ic)	Voltage (Vce)
TIP32	3 A	40 V
TIP32A	3 A	60 V
TIP32B	3 A	80 V
TIP32C	3 A	100 V
TIP42	6 A	40 V
TIP42A	6 A	60 V
TIP42B	6 A	80 V
TIP42C	6 A	100 V

Figure 16.9e.   High-power circuit using power FETs

A power MOSFET can be used to drive high-current loads from logic outputs by adding only one resistor to the circuit. (In some cases, even this resistor can be omitted and the output of the IC connected directly to the gate of the transistor.) Any power MOSFET (such as the IRF series devices) can be used in your application, since all they can control currents that, in some cases, rise to as high as 100 A!

Obs.: The standard power MOSFET is an N-channel unit. Although P-channel units exist, which drive a load when the output of the CMOS IC goes to the LOW logic level, they are not very common in practical applications. If you need to drive a load when the logic level is LOW, consider the use of an inverter stage.

Driving an SCR in a Low-Voltage DC Circuit

Sensitive SCRs such as devices in the 106 series (TIC106, MCR106, IR106, etc.) can be triggered from the low-power pulses produced in the output of a logic circuit. The TIC106 SCR can control loads up to 3 A. Note that the voltage drop across an SCR when in the conductive state is about 2 V.

Figure 16.9f.   Driving an SCR in a high-voltage AC circuit (half wave)

In AC circuits, sensitive SCRs such as devices in the 106 series can be triggered directly from the logic outputs as shown in this figure. In some cases, to bias the SCR to avoiding erratic triggering, the resistor R is required. In particular, with the TIC106, this resistor can range from 10 to 470 kΩ. It is important to observe that the low-voltage section of the circuit shares a common ground with the high-voltage section.

Figure 16.9g.   Driving high-current SCRs (NPN)

SCRs in the 106 series are very sensitive, needing less than 1 mA to be triggered. If we use high-power SCRs that need more current to be triggered in our application, this is the indicated circuit. This circuit is designed for SCRs of the TIC series, which need currents in the range between 10 and 100 mA to be triggered. In this application, the SCR is triggered when the output of the logic block goes to the HIGH logic level. The transistor is a BD135 or any other medium-power silicon

NPN transistor, and R is between 47 and 220 Ω. The supply voltage for this stage is between 6 and 15 V.

Figure 16.9h.   Driving a triac

In general, triacs need more current than SCRs to be triggered, and some cannot be connected directly to the output of a logic circuit. The transistor is a BD135, BD137, or any other medium-power NPN transistor. The value of resistor R is between 47 and 220 Ω, and it is chosen according to the sensitivity of the SCR and the voltage, +V. This voltage stays between 9 and 15 V.

Figure 16.9i.   Driving an IGBT

Isolated-gate bipolar transistors (IGBTs) are high-power devices used to control high-current loads in DC circuits.

Figure 16.9j.   Using an optocoupler, version 1

The previously discussed circuits in which devices powered from the AC power line are controlled from logic outputs had an important limitation: the presence of a common ground for the high-voltage stage and the logic circuit. This common ground can represent a shock hazard if any part of the low-power circuit is touched by the operator. A solution for triggering high-power devices from the AC power line involves the use of optocouplers. These devices incorporate an LED that acts on a sensor, thereby isolating the signal source from the receiver. Isolation voltages of more than 7,000 V are common in these devices.

Figure 16.9k.   Using an optocoupler, version 2

This circuit can be used to isolate blocks using logic elements. The value of resistor R depends on the power supply voltage per the following table:

Supply voltage	Resistor R
5 V	120 Ω
6 V	150 Ω
9 V	220 Ω
12 V	470 Ω
15 V	680 Ω

The values in the table are approximate and depend on the sensitivity of the phototransistor. Depending on the application, a tolerance range of 50 percent can be adopted.

Not shown in the figure.   Driving a triac with an optodiac (127 VAC)

The MOC3010 optodiac is formed by an infrared LED and a optodiac, a trigger element for triacs. This device is ideal for triggering triacs from CMOS (and other) logic outputs.

Power supply voltage	Resistor R
5 V	180 Ω
6 V	220 Ω
9 V	470 Ω
12 V	820 Ω
15 V	1k2 Ω

It is important to remind the reader that two other optocouplers exist in same family (MOC3011 and MOC3012) that are more sensitive, therefore needing less current to be triggered. If the triac doesn't trigger with the values of the resistors chosen using the table, you can reduce the values or use another optodiac.

Figure 16.9l. Driving a triac with an optodiac (220/240 VAC)

We use the MOC3020 optodiac (or more sensitive units such as the MOC3021 and MOC3022) in this configuration. R is chosen using the table given for the previous circuit.

LOGIC CIRCUIT INTERFACES

CMOS logic circuits can interface with other logic families, such as TTL, and even with operational amplifiers and comparators. In some cases, it is not possible to connect the output of one circuit directly to the input of the other; some kind of interface circuit is needed. Below, we show how to interface a CMOS circuit with other logic blocks and operational amplifiers (see Fig. 16.10).

Figure 16.10a. TTL to CMOS (5 V)

Figure 16.10b. TTL to CMOS (different voltages)

Figure 16.10c. TTL open collector to CMOS (different voltages)

Figure 16.10d. CMOS to TTL (5 V)

Figure 16.10e. CMOS to TTL (different voltages)

Figure 16.10f. CMOS to TTL using a CMOS buffer

Figure 16.10g. Operational amplifier to CMOS (same voltage)

Figure 16.10h. Operational amplifier to CMOS (different voltages)

Figure 16.10i. TTL to CMOS using an optocoupler

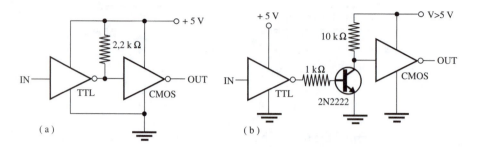

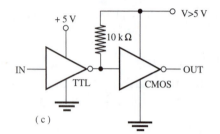

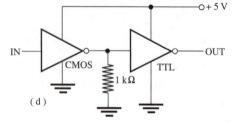

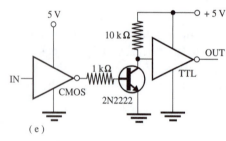

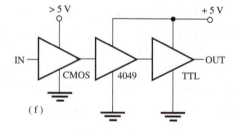

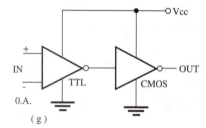

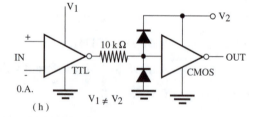

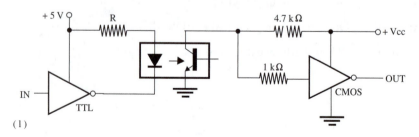

Figure 16.10

CHAPTER 17

Safety

THE IMPORTANCE OF SAFETY

You probably would never expect that your experiment with a mechatronic device would involve Asimov's laws of robotics (see Fig. 17.1). However, homemade and experimental projects that use common parts and are controlled by nonspecialized persons involve increased probability of an accident. This means that your experimental robot, mechatronic arm, or elevator can hurt you or other nearby people.

In simple words, this means that you are not free from the possibility of an accident—having a finger caught by a gear, cut by a gripper, or struck with an electric shock when touching a high-voltage source in your project.

Since this book is intended for the designer of all kinds and levels of mechatronic devices, and it includes high-power circuits that have the potential for a fatal accident, basic rules of safety are extremely important.

The robots and mechatronic devices that you can build may move among people and touch them. If you lose control of them, they can be dangerous and cause harm.

Isaac Asimov's Laws of Robotics

0. A robot may not injure humanity or, through inaction, allow humanity to come to harm.
1. A robot may not injure a human being, or through inaction allow a human being to come to harm, except where that would conflict with the 0'th law.
2. A robot must obey the orders given it by humans beings except where such orders would conflict with the first law.
3. A robot must protect its own existence as long such protection does not conflict with the first or second law.

Isaac Asimov, 1950

Figure 17.1

A heavy robot that collides against a child, or an experimental elevator that pinches the finger of a careless operator, are examples of accidents to be avoided when working in this field. Because this potential for accidents exists when working with mechatronic projects, this chapter is included to remind the reader of the importance of safety.

EXPERIMENTAL AND INDUSTRIAL ROBOTS

In terms of robots, the issue safety first became relevant when industrial robots increased in importance and number, as did the number of accidents involving maintenance personnel and workers. A document released by the National Institute for Occupational Safety and Health (NIOSH) offers safe maintenance guidelines for robotic workstations (http://www.cdc.gov/niosh/88-108.html). Although this document is primarily directed at people who work in the maintenance of industrial robot installations, the basic procedures given in its pages are valid for everyone who works with mechatronic devices and robots.

We need to remember, however, that there are basic differences between industrial robots and experimental ones that the reader can build with common parts, many of which are built at home or in the school lab. The robots used in industry and other real-world applications are very complex devices, often designed for single tasks such as soldering or welding workpieces in a car assembly line, placing objects in a predetermined position, moving pieces from one place to another, and so forth.

Taking a brief look in these machines and comparing them with our small experimental robots and other mechatronic experimental devices, we find three points where the differences are evident:

- The amount of power involved
- The degree of complexity
- The environment in which they operate

Power

Industrial robots are powerful units, and in many cases they can handle heavy pieces of metal or other materials, moving and placing them in specific positions. These machines are very dangerous because, if they operate in an uncontrolled mode, this power has the potential to cause physical harm or even fatal injuries. They can injure by impact, puncture, or pinch point closure, and they can drag someone into a sharp object or push a person into another machine.

Small experimental robots probably can't push you against a wall or throw you down a staircase with their arm movement, but they do have the potential to cause injuries. The limited power of an experimental robot usually limits its potential for severe injury, but you can't ignore the possibility of having a finger cut off or caught in the gears of the mechanism.

Complexity

Another factor to be considered in terms of safety is the degree of project complexity. The probability of an unintended device movement or an unexpected reaction when performing specific task increases with the complexity of the circuit and the number of functions it performs.

Many precautions are taken with industrial robots to avoid danger to people who work with them, and in particular to maintenance personnel. Among them are emergency switches that cut power in case of a malfunction and sensors that detect the presence of humans in dangerous locations. When working with experimental robots, the danger is not so significant, but you should consider measures that will avoid accidents in case of uncontrolled or unexpected operational states.

The Ambient

Industrial robots are highly specialized machines that operate in environments where only authorized or trained personnel are admitted. These people are trained to work in dangerous places and know what they have to do in case of emergency.

In general, however, experimental robots and mechatronic devices are not operated by individuals who have been trained in emergency procedures. In addition, in most cases, our devices operate among ordinary people, in rooms with children or students, or in your home or office environment.

Although the potential for fatal accidents is not as high as in the case of industrial robots, the fact that untrained people may be near the robot means that we should be doubly cautious and institute safety measures so as to avoid accidents.

SAFETY RULES

Accidents caused by a robot or mechatronic device derive from three main sources.

1. Mechanical hazards

   Accidents involving the mechanical parts of a robot or mechatronic device generally involve gears, wheels, grippers, and other moving parts that can cut, pierce, or crush parts of the human body. With the aim of avoiding these accidents, some basic rules are recommended:
   - Cover potentially hazardous moving parts to avoid contact with any part of your body. Gears, wheels, chains, and other parts must be covered. If you want to maintain visibility of the part, you can use transparent plastic, acrylic, or other such materials as covers.
   - Consider the use of sensors to detect abnormal operating conditions that may be indicative of an accident. For example, a stalled gear or wheel that increases the current drain of a motor or other device can reveal that something is caught in the mechanism; current sensing can be added to detect when a motor runs in an overloaded condition.

– When designing your project, avoid dangerous parts such as sharp points and cutting edges. If your robot is intended for competitions such as robot wars, think about the possible consequences of the weapons that you employ. In this case, it is a good idea to conduct combat inside protective barriers.

– Consider carefully the ambient in which your robot or mechatronic device will operate. Keep the robot away from objects that can be caught in its moving parts. Remove dangerous objects from the vicinity. If possible, operate the robot in a place that contains nothing but the objects that it is intended to manipulate or detect.

2. Electronic hazards

The electronic circuits used in our projects are not dangerous, as they are placed inside the device and, in general, are driven by low-power (and low-voltage) sources. However, some safety measures must be considered. The following are some safety rules that are indicated to protect users of robots and mechatronic devices from shock hazards and other dangers caused by the failure of electronic or electrical parts.

– Include emergency switches that will cut off power to the circuit if a problem occurs. (This measure is also useful to prevent accidents in which mechanical parts are involved.)

– Shield all parts in which high voltages are present. If possible, avoid the use of metallic parts or metal enclosures with high-voltage circuits.

– Include fuses or current-limiting circuits in all sensitive circuits to avoid problems in case of a short circuit.

– Consider the use of redundant protection in critical circuits, especially if the robot or mechatronic device will be operated by untrained people (children in particular).

– If your project includes self-defense capabilities or any other circuits that can cause injury, consider using redundant protection that will disable them in an emergency situation.

3. Chemical hazards

The use of chemical substances in any part of a robot or mechatronic device increases the probability of accidents. Substances inside a battery or fuel cell, gases produced by chemical cells or lead-acid batteries, substances used to produce special effects (smoke machines, fire extinguishers, etc.), and weapons (chemical attack) can cause accidents if improperly used. The basic rules to prevent accidents with chemical substances are as follows:

– Provide a secure exhaust channel for any gas produced inside the robot or mechatronic device.

– Be sure that any chemical product used to produce special effects cannot cause injury to persons in case of an accidental exposure to the eyes or skin.
– Keep a fire extinguisher nearby if any substance used in your robot can catch fire.
– Avoid the use of chemical-based special effects in closed rooms.

Artificial Intelligence

Circuits and software used in artificial intelligence are dangerous in the sense that they can control mechanisms that have a high potential accident risk. When running out of control, they can cause all of the problems discussed above.

The rules for preventing accidents caused by these circuits are the same as recommended previously. There is only one additional rule: take care to not design a device that can become more intelligent than you are.... You can lose control of it.

APPENDIX A

Mechatronics and the Internet

This chapter presents a list of useful Internet sites where the reader can find electronic components, kits, and literature. It also includes organizations and schools that are active in the fields of mechatronics, robotics, and artificial intelligence.

Aerotech Inc., Motion Control Division
Linear and rotary servomotors, software, and transducers
http://www.aerotechinc.com/mcdpro.html

All Electronics Corp.
Electronic Devices, kits, DC motors, etc.
http://www.allelectronics.com/

Allen Bradley
Industrial controls and automation
http://www.ab.com/

Americam Science & Surplus
Bearings, couplings, gears, springs, shafts, belts, pulleys, seals, etc.
http://sciplus.com/

Circuit Specialists Inc.
Electronic devices, robotics, test equipment, electronic chemicals, etc.
http://www.web-tronics.com/

Continental Hydraulics
Pumps, valves, etc.
http://www.continentalinc.ca/

Edmund Scientific
Optics, lens, mirrors, lasers, color filters, etc.
http://www.edsci.com/

Endveco Corp.
Accelerometers and other sensors
http://www.endevco.com/

Jameco Electronics
Electronic components, motors, relays, test and measurement equipment, etc.
http://www.jameco.com/

Ledex & Dormeyer Products
Linear solenoids and actuators
http://www.ledex.com/frames/indexProd.html

Lego
Mindstorms—blocks for robotics and mechatronics
http://mindstorms.lego.com/

MaxCim Motors, Inc.
Brushless DC motors and digital speed controls—high power
http://www.maxcim.com/

Measurement Computing Corp. (formerly Computer Boards)
Computer-based control and data acquisition
http://www.computerboards.com/

Microchip Technology Inc.
PIC microcontroller
http://www.microchip.com/

Mondo-Tronics Inc.
Distributor of SMAs, electric pistons, books, kits, and software
http://www.mondo.com/

Motion Online, Div. Sanyo Denki Co.
Servos, linear controls, multiaxis controls
http://www.MotionOnline.com/

National Instruments Corp.
Data acquisition, measurement, and automation
http://www.ni.com/

National Semiconductor Corp.
COP8 microcontroller
http://www.national.com

Newark Electronics
Electronic components, DC motors, etc.
http://www.newark.com/

Omega Engineering, Inc.
All types of sensors—distributor
http://www.omega.com/

Oriental Motor USA Corp.
Brushless, AC/DC motors, stepper motors, servo motors, etc.
http://www.omusa.com/index.html

Pacific Scientific
Motors, controllers, drivers, etc.
http://www.pacsci.com/

Parallax, Inc.
BASIC Stamp microcontroller
http://www.parallaxinc.com/

Parker Automation Technologies, Compumotor Division
Motion control products, literature, and software
http://www.compumotor.com/

Phoenix Contact
Data acquisition and control
http://www.iorail.com/

Radio Shack
Kits, electronic devices, DC motors, batteries, relays, solenoids, literature, etc.
http://www.radioshack.com/

Robot Store
Hobby robots and supplies—kits, assembled robots, literature, software, SMAs, PICs, electric pistons, etc.
http://www.robotstore.com/

Sensotec, Inc.
Sensors, accelerometers; pressure, oil, gas, torque, etc.
http://www.sensotec.com/index.html

Small Parts Inc.
Bearings, couplings, gears, springs, shafts, belts, pulleys, seals, etc.
http://www.smallparts.com/

Texas Instruments
DSPs—TMS320Cxxx
http://dspvillage.ti.com/docs/dspproducthome.jhtml

Victory Controls, LLC
Hydraulic and pneumatic controls and systems
http://www.victorycontrols.com/

Yaskawa Electric America, Inc.
Motion control, drives, and robotics
http://www.yaskawa.com/website/products/frproducts.htm